T0282427

Understanding the Bouguer Anomaly

Understanding the Bouguer Anomaly

A Gravimetry Puzzle

Roman Pašteka
Comenius University, Bratislava, Slovak Republic

Ján Mikuška
G-trend, s.r.o., Bratislava, Slovak Republic

Bruno Meurers
University of Vienna, Vienna, Austria

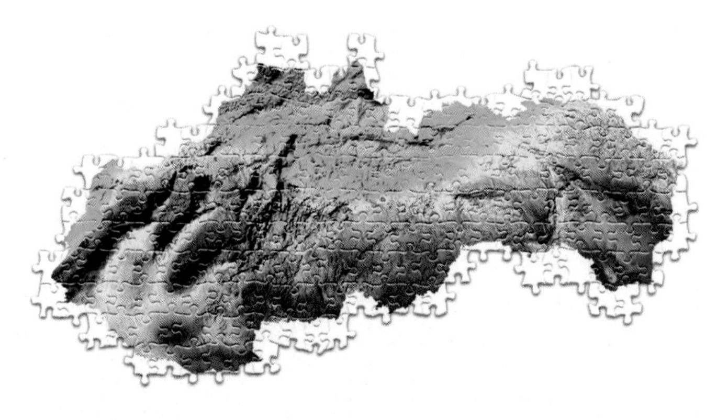

elsevier.com

Elsevier
Radarweg 29, PO Box 211, 1000 AE Amsterdam, Netherlands
The Boulevard, Langford Lane, Kidlington, Oxford OX5 1GB, United Kingdom
50 Hampshire Street, 5th Floor, Cambridge, MA 02139, United States

British Library Cataloguing-in-Publication Data
A catalogue record for this book is available from the British Library

Library of Congress Cataloging-in-Publication Data
A catalog record for this book is available from the Library of Congress

ISBN: 978-0-12-812913-5

Publisher: Candice G. Janco
Acquisition Editor: Marisa LaFleur
Editorial Project Manager: Marisa LaFleur
Production Project Manager: Stalin Viswanathan

Typeset by MPS Limited, Chennai, India

Chapter 7 National Gravimetric Database of the Slovak Republic 113

Pavol Zahorec, Roman Pašteka, Ján Mikuška, Viktória Szalaiová, Juraj Papčo, David Kušnirák, Jaroslava Pánisová, Martin Krajňák, Peter Vajda, Miroslav Bielik and Ivan Marušiak

7.1 Introduction 113
7.2 Compilation of Integrated Gravity Database 114
7.3 New Generation Bouguer Anomalies 118
7.4 New Linear Features Recognized in the Bouguer Anomaly Map 121
7.5 New Software for the Reconstruction of the Gravity Field from the Bouguer Anomaly Map 123
7.6 Conclusions 124
Acknowledgments 125
References 125

Conclusions 127

Roman Pašteka, Ján Mikuška and Bruno Meurers

LIST OF CONTRIBUTORS

Miroslav Bielik
Comenius University, Bratislava, Slovak Republic

Hans-Jürgen Götze
Christian Albrechts University, Kiel, Germany

Roland Karcol
Comenius University, Bratislava, Slovak Republic; Slovak Academy of Sciences, Bratislava, Slovak Republic

Martin Krajňák
Comenius University, Bratislava, Slovak Republic

David Kušnirák
Comenius University, Bratislava, Slovak Republic

Ivan Marušiak
G-trend, s.r.o., Bratislava, Slovak Republic

Bruno Meurers
University of Vienna, Vienna, Austria

Ján Mikuška
G-trend, s.r.o., Bratislava, Slovak Republic

Jaroslava Pánisová
Slovak Academy of Sciences, Bratislava, Slovak Republic

Juraj Papčo
Slovak University of Technology, Bratislava, Slovak Republic

Roman Pašteka
Comenius University, Bratislava, Slovak Republic

Viktória Szalaiová
Geocomplex, a.s., Bratislava, Slovak Republic

Wolfgang Szwillus
Christian Albrechts University, Kiel, Germany

Peter Vajda
Slovak Academy of Sciences, Bratislava, Slovak Republic

Pavol Zahorec
Slovak Academy of Sciences, Banská Bystrica, Slovak Republic

CHAPTER 1

Introduction

Roman Pašteka[1], Ján Mikuška[2] and Bruno Meurers[3]
[1]Comenius University, Bratislava, Slovak Republic [2]G-trend, s.r.o., Bratislava, Slovak Republic
[3]University of Vienna, Vienna, Austria

Measurement of gravity acceleration (together with estimation of the gravitational constant) was among the first physical experiments, and until today it retains high importance—in areas of theoretical physics to physical geodesy and applied geophysics. As it is typical in most physical experiments, the final result is dependent on two fundamental aspects—on the measurement itself and the results after generally accepted processing of the acquired experimental data. From the viewpoint of applied geophysics (which we define as the study of the subsurface structure of the Earth using physical measurements), the processing of gravity data plays a very important role. The reason for that is that gravity acceleration values themselves are strongly dependent on various nongeological phenomena (latitude, height, influence of topography), which mask the manifestation of subsurface density inhomogeneities in the measured gravity field. Therefore the main output from gravity data processing, in our efforts to make the data suitable for geophysical interpretation, is always a "gravity anomaly." "The gravity anomaly is defined as the difference between the observed and the theoretical or predicted vertical acceleration of gravity" (Hinze et al., 2013, p. 143). We find a very similar definition (even more concrete) in LaFehr and Nabighian (2012, p. 90): "A gravity anomaly is defined as the difference between measured gravity and theoretical gravity based on a defined earth model." Through this process of elimination of unwanted, nongeological, gravity effects we are able to interpret anomalous gravity with the aim to recognize and describe the geological sources of this field. At this point it is good to emphasize that the removal of these unwanted nongeological effects is accomplished using a mathematical description of the synthetic gravity field for the defined Earth model.

Understanding the Bouguer Anomaly. DOI: http://dx.doi.org/10.1016/B978-0-12-812913-5.00011-7

Several different definitions exist for describing various gravity anomalies. The most important are the so-called free-air and Bouguer gravity anomalies. Free-air anomalies are, in general, not suitable for geological interpretations, owing to their strong correlation with topography. Several authors use the term Faye anomaly as a synonym for free-air gravity anomaly (e.g., Pick et al., 1973; Telford et al., 1990), while many others distinguish among these two kinds of anomalies (for details, see Table 1.1). The Bouguer anomaly (BA) is a very important tool for geological interpretation. It was named after Pierre Bouguer (1698–1758), a well-known French astronomer, geodesist, and physicist, who was one of the pioneers in gravimetry. This publication is focused on several selected topics from the area of BA evaluation. A general definition of BA can be written in the form of a simple equation (e.g., LaFehr, 1991):

$$\mathrm{BA} = g_{\mathrm{obs}} - (g_0 + g_{\mathrm{f}} + g_{\mathrm{B}})$$

where g_0 is the latitude-dependent theoretical value of gravity at the vertical reference datum, g_{f} is the elevation-dependent free-air term, and g_{B} is the elevation- and topography-dependent Bouguer term (later in the book we will discuss more precise equations, see Chapter 2: The Physical Meaning of Bouguer Anomalies—General Aspects Revisited and Chapter 7: National Gravimetric Database of the Slovak Republic). It is important to mention here that during the removal of the g_0 term, the effect of centrifugal acceleration (due to the rotation of the Earth) is removed from the processed gravity acceleration value and the final BA value is consequently understood as a component of gravitational acceleration.

When discussing the various aspects of the BA evaluation in the past, readers can come across numerous discrepancies in its definition expressed either in textbooks or monographs (sometimes even including modern texts), which have been published within the last 25 years. In almost all contributions to this topic it is clearly stated that measured gravity has to be processed in the way that unwanted non-geological influences on gravity acceleration must be removed, but the way it is done and mainly understood is, in general, not unified. Discrepancies are mainly in the area of:

1. the understanding of the position of BA evaluation (the problem of a possible relocation of processed gravity values to some different reference height datum),

Table 1.1 Summary of Basic Terms and Aspects of Bouguer Anomaly Definition in Various Textbooks and Monographs

References	Used Terms in BA Evaluation	Form of BA Evaluation	Title of BA	Comments
Nettleton (1940, pp. 16–22, 51–62, 133–134)	Corrections (but gravity is reduced)	Equation in a symbolic form (equation in p. 133)	Bouguer anomaly	(1) Bouguer gravity is the theoretical gravity (normal field with following corrections) and (2) terrain effect term T is finally added in BA evaluation
Grant and West (1965, pp. 235–243)	Corrections (but gravity is reduced)	Do not give a final equation in a symbolic form, only partial steps are given	Bouguer gravity or Bouguer gravity anomaly	(1) Bouguer gravity is the title for the final anomalous field and (2) authors mention "free-air anomaly" or "free-air disturbance"
Torge (1989, p. 100)	Reductions	Equation in a symbolic form (Eq. 4.36)	Bouguer anomaly Δg_B	Topographic reduction term δg_{Top} is subtracted from free-air anomalies
Telford et al. (1990, pp. 11–15)	Corrections (but gravity is reduced)	Equation in a symbolic form (Eq. 2.28)	Bouguer anomaly g_B	They explicitly define the Bouguer anomaly value for the station
Blakely (1996, pp. 136–150)	Corrections	Equation in a symbolic form (Eq. 7.17)	Bouguer anomaly Δg_{SB} and Δg_{CB}	Distinguishes between simple and complete BA
Jacoby and Smilde (2009, pp. 151–166)	Reductions (authors avoid the term correction)	Equation in a symbolic form (equations in p. 165)	Bouguer anomaly BA or δBA	(1) Distinguish between free-air and Faye anomaly (2) distinguish between simple and complete BA
LaFehr and Nabighian (2012, pp. 81–100)	Corrections (but gravity is reduced)	Equation in a symbolic form (equation in p. 91)	Bouguer anomaly g_b	Distinguish between simple and complete BA
Hinze et al. (2013, pp. 122–155)	Corrections	Equation in a symbolic form (Eqs. 6.37 and 6.38)	Bouguer anomaly g_{SBA} and g_{CBA}	Distinguish between simple and complete BA
Long and Kaufmann (2013, pp. 17–39)	Corrections (but gravity is reduced)	Equation in a symbolic form (Eq. 2.40)	Bouguer anomaly Δg_B	(1) Distinguish between simple and complete BA (2) distinguish between free-air and Faye anomaly
Fairhead (2016)	Corrections	Equation in a symbolic form (in various modifications) (Chapter 2.1.3)	Bouguer anomaly	Distinguishes between simple and complete BA

2. technical aspects (e.g., different height systems—use of ellipsoidal vs orthometric heights) and selection of all evaluated corrections (e.g., atmospheric correction),
3. improper, or at least, inconsistent use of terminology and notation in BA evaluation (corrections vs reductions, anomaly vs disturbance, mathematical expression of BA equation).

Thanks to important discussions among US experts in the beginning of this millennium (Hinze et al., 2005; Li et al., 2006; Keller et al., 2006; together with other important papers—e.g., LaFehr, 1991; Chapin, 1996; Talwani, 1998; LaFehr, 1998; Li and Götze, 2001), which have been connected with the setting of new BA calculation standards in North America, many aspects have been made clearer and defined in a more correct way—from the scope of applied geophysics. Textbooks and monographs, published during the last several years (e.g., Jacoby and Smilde, 2009; LaFehr and Nabighian, 2012; Hinze et al., 2013; Long and Kaufmann, 2013; Fairhead, 2016) were written from these aspects in a much better way and BA evaluation is defined correctly from the viewpoint of applied gravimetry. In Hinze et al. (2013, pp. 145–151) readers can find several modifications of BA evaluation for different acquisition situations (a good overview for a planar approximation is also given in the NGA Report, 2008, p. 4).

One of the most important aspects of BA evaluation is the fact that the goal of the applied processing steps (reductions/corrections) is not to "reduce" measured gravity values (as explained in several older textbooks, e.g., Grant and West, 1965; Telford et al., 1990), but rather, they are used during "tuning" or improvement of the theoretical gravity field. The goal of this tuning is to obtain precise theoretical values of gravity acceleration of a reference Earth model, which are then removed from the measured (observed) values, producing an anomaly field. In the classical work of Hayford and Bowie (1912) we can find on pages 72 and 73 the following statement: "Usually corrections are applied to the observed values of the intensity of gravity to reduce them to sea level and to correct for the supposed influence of topography. In this publication the corrections are applied to the theoretical value of the intensity of gravity at the sea level to obtain the theoretical value at the station, a value which is directly comparable with the observed value. This seems to the authors to be more logical method and more conductive to clear thinking than the usual method."

This concept was also very well understood by Nettleton (1940)—who entitled the evaluated and subtracted theoretical fields as *Bouguer gravity* and the final received anomalous field as *Bouguer anomaly* (see also Table 1.1). Unfortunately, many authors later on have understood these two quantities as synonyms. Fortunately, in the actual textbooks mentioned about, this aspect is presented in the correct way from the viewpoint of applied gravimetry.

The next most important aspect of BA evaluation is the fact that in applied gravimetry we do not move or relocate the processed gravity value to some reference datum (we can find this incorrect explanation in several older textbooks and papers, e.g., Jung, 1961, pp. 82, 83; Dobrin, 1976, p. 417), but, instead, the positions of the BA values "reside" at the acquisition points. This is well explained in Jacoby and Smilde (2009, p. 155): "Note that reductions do not physically move the observations to another level – as incorrectly expressed frequently. Reductions compare the observation with normal values by estimating the latter at observation locations ... Anomalies are determined at the observation points, not at the reference level." Another clear statement to this point can be found in LaFehr and Nabighian (2012, pp. 86, 87): "Let us be clear that by making this reduction, we are not reducing the data to a datum, i.e., obtaining at a fictitious station on the datum what we would have measured if we had been able to do so." Here it is important to mention the short note from Ervin (1977), where he point to the absurdity of such kind of fictive relocation of processed values. On the other hand, we can still find in actual geophysical monographs (e.g., Mallick et al., 2012), this typical old-fashioned interpretation coming from physical geodesy (where experts need to know the gravity acceleration at the geoid datum), but in the majority of up-to-date geophysical textbooks this aspect is explained correctly.

The terminology used in BA evaluation belongs also to the problems discussed, especially the meaning of the terms reduction and correction. These two terms express the same steps in the processing of gravity data, but sometimes they can be understood in different ways. For example, Nettleton (1940) used these terms as synonyms, while some other authors write about reduction of gravity values—but the partial steps are titled as corrections (for an overview, see Table 1.1, second column). The majority of authors of this monograph chapters prefer the use of the term correction, because the term reduction can indicate the sense that something could be moved in the downward direction (reduced), which could contribute to misunderstanding the physical location of BA values (as discussed earlier). It would be very helpful to have a unification of such important terms in gravimetry, but also here the most important principle is to understand the real content of the used terms.

The final comment to the terminology in this introduction is focused on the problem of "anomaly" versus "disturbance." In physical geodesy experts are distinguishing between these two terms—an anomaly is a difference between quantities known at different points, whereas a disturbance is valid for this kind of calculation at the same point. So, from this viewpoint, BA, as we understand it in applied gravimetry, should be called a "disturbance" (see more details in Chapter 2: The Physical Meaning of Bouguer Anomalies—General Aspects Revisited), when the vertical datum for the used heights is the ellipsoid. But we agree (i.e., the majority of editors of this book) with the statement by Hinze et al. (2005, p. J31) that "The geophysical community is largely unaware of the term disturbance as used by geodesists, which will add further confusion if the term is used." So, we suggest to continue using the term "anomaly" in the case of BA definition.

The motivation for preparing this book was the excellent experience which editors and authors of individual chapters have made during the 2-day workshop with the title "Bouguer anomaly - what kind of puzzle it is?," which was held in September 2014 in Bratislava. It was organized in the frame of a scientific project "Bouguer anomalies of new generation and the gravimetrical model of Western Carpathians." This project was supported by the Slovak Research and Development Agency and conducted by the Department of Applied Geophysics, Comenius University in Bratislava, together with the Geophysical Institute of the Slovak Academy of Sciences and the companies G-trend s.r.o. and Geocomplex Inc. (more information about the project outputs can be found in Chapter 7: National Gravimetric Database of the Slovak Republic). The workshop was attended by several well-known experts from the gravimetrical branch (in alphabetic order: Benedek, J., Bielik, M., Braitenberg, C., Čunderlík, R., Dérerová, J., Ebbing, J., Gabriel, G., Hronček, S., Karcol, R., Li, X.,

Meurers, B., Mikuška, J., Mrlina, J., Pánisová, J., Papp, G., Pašteka, R., Pohánka, V., Ruess, D., Scücs, E., Smilde, P., Szwillus, W., Švancara, J., Vajda, P., and Zahorec, P.). During the workshop we had excellent, thorough discussions on the topic of BA evaluation problems. This workshop can be understood also as a "continuator" of local gravimetrical conferences AGK ("Alpen-Gravimetrie-Kolloqium"), which had been organized mostly by colleagues from Austria, periodically from the mid-1980s until 2006, and have been hosted by several central European institutions and universities. After the end of the last mentioned Bratislava workshop in 2014, several presenters carried out an intensive discussion regarding the possibility of publishing several of the presented topics in the form of a widely accessible publication. This book is the result of this discussion.

It is organized logically in eight chapters, including Introduction and Conclusion, where the topics are ordered from aspects of BA definition, through history of BA evaluation, normal field (an alternative) evaluation and calculating terrain corrections, finishing with an example of the national gravity database organization and recalculation.

In the second chapter titled "The Physical Meaning of Bouguer Anomalies—General Aspects Revisited" Bruno Meurers picks up the approach of composing synthetically all gravitational effects contributing to the gravity observed at the Earth's surface. This concept enables us to precisely specify the physical meaning of Bouguer anomalies obtained under certain assumptions (e.g., reference surface). The chapter also tries to assess the errors, which inherently occur due to specific assumptions made in calculating the BA in practice. This is focused on the problem of the scalar representation, the geophysical indirect effect, the truncation of the mass correction area, and the normal gravity calculation in areas with negative ellipsoidal heights.

The third chapter titled "Some Remarks on the Early History of the Bouguer Anomaly" from Mikuška et al. investigates in detail the historical aspects of several important parts of the BA evaluation. For instance, one can find that the term "Bouguer reduction" is most likely accreditable to Helmert. This term was used in the context of the popular geodetic concept of reducing gravity from the Earth surface to sea level, although Bouguer himself had never reduced gravity values or pendulum lengths in such a manner. Authors confirm that the foundations of the so-called "Bouguer anomaly" and the procedures applied to the measured gravity, which are required by the gravity method of applied geophysics, can be tracked back to originate from the famous book of Pierre Bouguer published in 1749. Among others, various authors call attention especially to one specific historical artefact, namely the misunderstanding associated with the gravity data reductions to sea level in applied geophysics which, although geophysically unacceptable, has withstood the ravages of time and can be found in the literature even in the 21st century.

In the fourth chapter titled "Normal Earth Gravity Field Versus Gravity Effect of Layered Ellipsoidal Model" Karcol et al. discuss the question of whether the present-day normal gravity field calculation is suitable from the aspect of applied gravimetry. Today the normal field or theoretical gravity is on a regular basis strictly related to a model of rotational biaxial ellipsoid with one important property—namely, the constant potential on its surface. This condition, however, would require a specific density distribution within the model. Treating the Earth atmosphere and topography is another issue. These masses are simply moved into the interior of such ellipsoid and "dissolved" there. In short, this approach does not fit the geophysical reality very well. Instead, in this chapter, the authors attempt to calculate the normal field as the gravity effect of a layered rotational ellipsoid, namely the gravitational effect of a suitable set of homeoid shells bounded by two similar ellipsoids having a constant ratio of axes, plus the centrifugal component. The authors believe that this approach will fit the structure and density distribution within the Earth much better than existing models. They analyze and present the primary differences between the two approaches, and also discuss the relation of the normal field to the free-air correction.

In the frame of the fifth chapter titled "Numerical Calculation of Terrain Correction Within the Bouguer Anomaly Evaluation" Zahorec et al. describe properties of numerical evaluation of the terrain corrections by means of a recently developed code Toposk (developed in order to recalculate the terrain corrections of the unified gravity database of the Slovak Republic). The program is designed primarily for calculating the gravitational effect of topographic masses. Terrain corrections are then derived from these effects. Its application is not restricted to the territory of Slovakia, but can be used worldwide. It allows using interpolated heights for the calculation points in order to reduce the errors resulting from the elevation model inaccuracy. The choice of arbitrary subzone divisions, as well as multiple options for coordinate systems, is allowed. By default the program uses the following zone divisions: inner zone T1 (to a distance of 250 m from the calculation point), intermediate zone T2 (250–5240 m), and outer zones T31 (5240–28,800 m) and T32 (28,800–166,730 m). The computing algorithm was tested on several synthetic models, giving satisfactory results when compared with analytically evaluated values.

In the sixth chapter titled "Efficient Mass Correction Using an Adaptive Method" Szwillus and Götze focus their attention on an effective way of numerical evaluation of topographic effects. These calculations are always very time-consuming. To increase the computational efficiency, the authors present an adaptive approach that enables users to perform large-scale or even globally complete correction. The central idea is to use only that part of the topography information that is actually relevant for the gravitational effects. Numerical analysis of the synthetic model shows that this adaptive approach is expected to perform well as long as its parameters are chosen according to the geostatistical properties of topography in the studied area. Real-world tests of a high-resolution topography dataset from the Himalayan Plateau demonstrate the reliability and efficiency of the adaptive approach.

In the seventh chapter "National Gravimetric Database of the Slovak Republic" Zahorec et al. present a case study of BA calculation for a large database (from the entire country). The results of the Slovak gravimetric database compilation, with actually more than 320,000 observation points, are presented (realized in the scope of the mentioned project "Bouguer anomalies of new generation and the gravimetrical model of Western Carpathians"). Gravity data were collected for more than 50 years, which yields a very heterogeneous dataset, with large variations in the station coverage and processing methods. The regional gravimetric database (more than 212,000 points) was resumed in 2001. The compilation discussed herein (with more than 107,000 detailed gravity measurements added to the original database) was made during the period 2011–14. A rigorous quality control process and complete recalculation of the Bouguer anomalies is presented. The primary focus of this project was on a proper recalculation of the terrain corrections. A new software solution for reconstruction of the gravity acceleration values from the BA map (program CBA2G_SK) was developed for geodetic applications.

All chapters of this book have been peer-reviewed by several independent experts. Among them we would like to thank to following external reviewers:

John Bain, Bain Geophysical Services, Houston, TX, United States,
Allen Cogbill, Geophysical Software, Inc., Los Alamos, NM, United States,
Serguei Goussev, Exploration Consultant, Vancouver, BC, Canada,
Pavel Novák, University of West Bohemia, Pilsen, Czech Republic,
Jan Mrlina, Academy of Sciences, Prague, Czech Republic.

At the end of this introductory part, we would like to express our hope that this book will positively contribute to the scientific development in the area of BA evaluation, with the aim to improve further geological interpretation of gravimetrical datasets. We would like to express our grateful thanks to our colleagues, who helped us with the organization of the mentioned workshop, mainly to Barbora Šimonová, Martin Krajňák, and Marián Bošanský. We are also thankful to the companies who have supported our efforts from a technical point of view, mainly to Proxima R&D and G-trend s.r.o. Special thanks go to John Bain for his support and many suggestions, which have helped to improve the scientific content of the published material in this book. Finally, we would like to thank the support from the publishing house—especially to Marisa LaFleur, Hillary Carr, Stalin Viswanathan, and Rakesh Venkatesan.

REFERENCES

Blakely, R.J., 1996. Potential Theory in Gravity and Magnetic Applications. Cambridge University Press, Cambridge, 441 p.

Chapin, D.A., 1996. The theory of the Bouguer gravity anomaly: a tutorial. Leading Edge 15 (5), 361–363.

Dobrin, M.B., 1976. Introduction to Geophysical Prospecting. McGraw-Hill, New York, NY, 630 p.

Ervin, C.P., 1977. Theory of the Bouguer anomaly. Short note. Geophysics 42, 1468.

Fairhead, J.D., 2016. Advances in Gravity and Magnetic Processing and Interpretation. EAGE Publications bv, DB Houten, 352 p.

Grant, F.S., West, G.F., 1965. Interpretation Theory in Applied Geophysics. McGraw-Hill, New York, NY, 584 p.

Hayford, J.F., Bowie, W., 1912. The Effect of Topography and Isostatic Compensation upon the Intensity of Gravity. U.S. Coast and Geodetic Survey, Special Publication No. 10, 132 p.

Hinze, W.J., Aiken, C., Brozena, J., Coakley, B., Dater, D., Flanagan, G., et al., 2005. New standards for reducing gravity data: the North American gravity database. Geophysics 70, J25–J32.

Hinze, W.J., Von Frese, R.R.B., Saad, A.H., 2013. Gravity and Magnetic Exploration. Cambridge University Press, Cambridge, 512 p.

Jacoby, W., Smilde, P.L., 2009. Gravity Interpretation: Fundamentals and Application of Gravity Inversion and Geological Interpretation. Springer, Berlin, 395 p.

Jung, K., 1961. Schwerkraftverfahren in der angewandten Geophysik. Geest&Portig K.-G348 p. (in German).

Keller, G.R., Hildebrand, T.G., Hinze, W.J., Li X., Ravat, D., Webring M., 2006. The quest for the perfect gravity anomaly: Part 2—Mass effects and anomaly inversion. SEG expanded abstract, New Orleans, SEG Annual Meeting, pp. 864–868.

LaFehr, T.R., 1991. Standardization in gravity reduction. Geophysics 56, 1170–1178.

LaFehr, T.R., 1998. On Talwani's "Errors in the total Bouguer reduction". Geophysics 63, 1131–1136.

LaFehr, T.R., Nabighian, M.N., 2012. Fundamentals of Gravity Exploration. SEG, Tulsa, 218 p.

Li, X., Götze, H.-J., 2001. Ellipsoid, geoid, gravity, geodesy and geophysics. Geophysics 66, 1660–1668.

Li X., Hinze W.J., Ravat D., 2006. The quest for the perfect gravity anomaly: Part 1—New calculation standards. SEG expanded abstract, New Orleans, SEG Annual Meeting, pp. 859–863.

Long, L.T., Kaufmann, R.D., 2013. Acquisition and Analysis of Terrestrial Gravity Data. Cambridge University Press, Cambridge, 171 p.

Mallick, K., Vashanti, A., Sharma, K.K., 2012. Bouguer Gravity Regional and Residual Separation: Application to Geology and Environment. Springer, New York, NY, 288 p.

Nettleton, L.L., 1940. Geophysical Prospecting for Oil. McGraw-Hill, New York, NY, 444 p.

NGA Report, 2008. Gravity station data format and anomaly computations. National Geospatial-Intelligence Agency, Internal Report (manuscript), 7 p.

Pick, M., Pícha, J., Vyskočil, V., 1973. Theory of the Earth's Gravity Field. Elsevier, Amsterdam, 538 p.

Talwani, M., 1998. Errors in the total Bouguer reduction. Geophysics 63, 1125–1130.

Telford, W.M., Geldart, L.P., Sheriff, R.E., 1990. Applied Geophysics. Cambridge University Press, Cambridge, 744 p.

Torge, W., 1989. Gravimetry. Walter de Gruyter, Berlin, 465 p.

CHAPTER 2

The Physical Meaning of Bouguer Anomalies—General Aspects Revisited

Bruno Meurers
University of Vienna, Vienna, Austria

2.1 INTRODUCTION

Interpreting the physical meaning of the Bouguer anomaly is an old problem associated with various assumptions and simplifications made for calculating the height and mass correction terms. It is also related to the purpose the anomaly is used for in different disciplines. For example, for solving the Stokes boundary value problem of physical geodesy we need to know the gravity anomaly at the geoid. In geophysics the goal is finding the gravitational response of subsurface masses on a given reference surface. Ideally, this should be a horizontal plane or a sphere because such reference surfaces simplify processing steps like field continuation or other field transformations applied in the interpretation. But 50 years ago Naudy and Neumann (1965) or Tsuboi (1965) emphasized that the Bouguer anomaly does not represent the gravity effect of anomalous density on a horizontal reference level. Tsuboi (1965) distinguished between station Bouguer anomaly and true Bouguer anomaly, clarifying that the Bouguer gravity is always related to points at the irregular observation surface and not to any other simple reference surface. Nevertheless, even in classical textbooks this fact is not emphasized accordingly (LaFehr, 1991). Another issue in this discussion is the height system used for calculating the normal gravity at a gravity station. Several authors showed that ellipsoidal heights rather than orthometric or normal heights should be used in order to keep a clear concept (e.g., Vogel, 1982; Meurers, 1992). A comprehensive presentation of all problems related to the normal gravity correction has been given by Li and Götze (2001). Hackney and Featherstone (2003) or Vajda et al. (2006) discussed this topic from the perspective of physical geodesy in terms of gravity anomaly and gravity disturbance in order to eliminate the misunderstanding

Understanding the Bouguer Anomaly. DOI: http://dx.doi.org/10.1016/B978-0-12-812913-5.00001-4

between geodesists and geophysicists. Regarding the mass corrections we have to deal with basic assumptions or limitations with respect to density, geometrical approximation of sources, and truncation effects. Hinze et al. (2005) provided a proposal for standard processing of gravity data taking the discussed aspects into account.

The basic assumptions and simplifications in the practical calculation of the Bouguer gravity are closely related to its error budget. The geophysicist has to know these errors for interpreting the anomalies correctly. Using a synthetic concept (e.g., Vogel, 1982) is an appropriate approach to discuss the physical meaning of the Bouguer anomaly and to assess the errors introduced.

Physical geodesy determines the geoid based on observations of gravity $\mathbf{g}$ at the Earth surface. The geoid is closely related to the disturbing potential T leading to the fundamental equation of physical geodesy (Eq. 2.1). It describes the general boundary value problem where Δg is the scalar representation of the gravity anomaly vector $\boldsymbol{\Delta}\mathbf{g}$ in a classical sense with γ denoting the normal gravity vector of the reference ellipsoid:

$$T(P_0) = W(P_0) - U(P_0) \quad \text{disturbing potential}$$

$$\boldsymbol{\Delta}\mathbf{g} = \mathbf{g}(P_0) - \gamma(P_E) \quad \text{gravity anomaly}$$

$$\Delta g = -\left(\frac{\partial T}{\partial n}\right)_{P_0} + \frac{1}{\gamma(P_E)}\left(\frac{\partial \gamma}{\partial n_E}\right)_{P_E} T(P_0) \tag{2.1}$$

Eq. (2.1) results from linearization and combines the derivatives in direction of the plumb lines of the respective gravity field. In geophysics, we are looking for the geometry and the density contrast of anomalous subsurface sources with respect to a reference density model. The definition of the Bouguer anomaly **BA** (Eq. 2.2b) implies using the gravity disturbance vector $\boldsymbol{\delta}\mathbf{g}$ (Eq. 2.2a):

$$\boldsymbol{\delta}\mathbf{g}(P) = \mathbf{g}(P) - \gamma(P) \tag{2.2a}$$

$$\mathbf{BA}(P) = \mathbf{g}(P) - \underbrace{\gamma(P_E)}_{\gamma_0} - \int_{P_E}^{P} \boldsymbol{\Gamma}(\mathbf{s}) \cdot d\mathbf{s} - \mathbf{g}_M(P) = \boldsymbol{\delta}\mathbf{g}(P) - \mathbf{g}_M(P) \tag{2.2b}$$

where $\mathbf{g}_M$ is the gravitational attraction vector caused by all masses above the reference surface including the atmosphere. That is the reason why some authors suggest calling it "topographically corrected gravity disturbance" (e.g., Vajda et al., 2006). $\boldsymbol{\Gamma}$ denotes the gradient tensor of normal gravity, and $d\mathbf{s}$ is an infinitesimally small line element along the plumb line. P_E is the intersection of the plumb line passing P and the ellipsoid. $\gamma(P_E) = \gamma_0$ is the normal gravity on the reference ellipsoid at P_E. Neglecting the deviation of the plumb line from the ellipsoidal normal simplifies Eq. (2.2b):

$$\mathbf{BA}(P) = \mathbf{g}(P) - \underbrace{\gamma(P_E)}_{\gamma_0} - \int_0^{z(P)} \frac{\gamma(z)}{\partial z} dz - \mathbf{g}_M(P) = \delta\mathbf{g}(P) - \mathbf{g}_M(P) \quad (2.2c)$$

where $z(P)$ is the station elevation in the chosen height system and P_E is now the projection of P along the ellipsoid normal (Fig. 2.1). The physical meaning of Eq. (2.2c) depends on the height system applied.

Composing the gravity $\mathbf{g}$ observed at point P on topography synthetically, we have to consider $\mathbf{g}(P)$ as

$$\mathbf{g}(P) = \gamma(P_E) + \int_0^{z=N} \frac{\gamma(z)}{\partial z} dz + \int_{z=N}^{z=N+H} \frac{\gamma(z)}{\partial z} dz + \mathbf{g}_S(P_E) + \int_0^{z=N} \frac{\mathbf{g}_S(z)}{\partial z} dz + \int_{z=N}^{z=N+H} \frac{\mathbf{g}_S(z)}{\partial z} dz + \mathbf{g}_T(P) + \mathbf{g}_G(P) + \mathbf{g}_A(P) \quad (2.3)$$

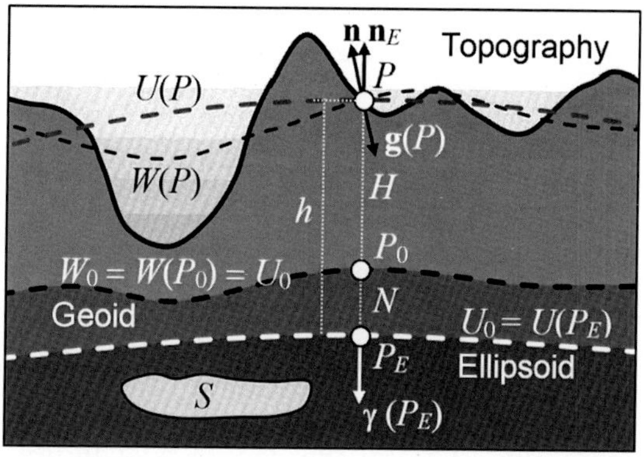

Figure 2.1 Definition of the Bouguer anomaly. Dashed lines show intersections with the equipotential surfaces of the gravity potential W *and the normal gravity potential* U, *respectively. Per definition, the gravity potential* W_0 *on the geoid equals to the normal gravity potential* U_0 *on the reference ellipsoid.* P *marks an observation point at topography. Its projections onto the geoid and the reference ellipsoid are denoted by* P_0 *and* P_E, *respectively.* $\boldsymbol{n}$ *and* $\boldsymbol{n}_E$ *are unit vectors aligned to the surface normal of the corresponding equipotential surfaces.* h *and* H *represent the ellipsoidal and orthometric height, respectively of* P. N *denotes the geoid undulation. The light-gray shaded area S stands exemplarily for all density anomalies inside and outside the ellipsoid.*

The total gravitational attraction vector $\mathbf{g}_M$ is composed of $\mathbf{g}_T$, $\mathbf{g}_G$, and $\mathbf{g}_A$ denoting the contribution of different density domains indicated by the subscripts T (mass between topography and geoid), G (mass between geoid and reference ellipsoid), and A (atmosphere). h describes the ellipsoidal height of P, H its orthometric height, and N the geoid undulation. The latter can be replaced by the normal height and the height anomaly if the normal height system is preferred. $\mathbf{g}_S$ represents the total gravitational attraction vector caused by all anomalous density sources inside and outside the ellipsoid (exemplarily shown by the *light-gray area* S in Fig. 2.1). In this context, anomalous density means the difference between true density and the density distribution of the reference ellipsoid or the model used for mass corrections. From theoretical point of view, we have to consider that the reference ellipsoid does not tell anything about the internal density structure because it is fully defined by only four Stokes' constants. However, Moritz (1990) has shown that structures like those of the preliminary Earth model have a gravity distribution close to the normal gravity. This permits using the normal gravity as a reference. Today there is no need to apply simplified formulae to calculate the normal gravity at the observation site. Either closed formulae including atmospheric correction should be used instead (e.g., Li and Götze, 2001) or second-order Taylor series representations as proposed by Wenzel (1985).

The physical meaning of the Bouguer anomaly is directly visible after inserting Eq. (2.3) into (2.2c), provided ellipsoidal heights are used:

$$\mathbf{BA}(P) = \mathbf{g}_S(P_E) + \int_0^h \frac{\mathbf{g}_S(z)}{\partial z} dz = \mathbf{g}_S(P) \tag{2.4}$$

Contrarily, if orthometric heights are used, we get:

$$\mathbf{BA}(P) = \mathbf{g}_S(P) + \int_0^N \frac{\mathbf{g}_S(z)}{\partial z} dz + \mathbf{g}_G(P) \tag{2.5}$$

which differs from Eq. (2.4) by the geophysical indirect effect (GIE; Hackney and Featherstone, 2003). A similar equation holds true in case of normal heights. Only the Bouguer anomaly evaluated in an ellipsoidal height system is identical with the gravitational attraction vector $\mathbf{g}_S(P)$ of unknown sources. Therefore the ellipsoidal height system should be preferred.

The synthetic concept (Eqs. 2.4 and 2.5) easily permits explaining the gravity effects left in the Bouguer anomaly depending on the basic assumptions made and on the definition of the density domains in Eq. (2.3). It also tells us how to proceed with interpreting the anomaly in the chosen three-dimensional (3D) model space. This even holds if we do not know exactly the density distributions assumed for the mass corrections.

The goal of this chapter is reviewing the problem trying to assess a few specific error sources quantitatively:

- Scalar versus vector representation
- GIE
- Mass correction (truncation effect)
- Normal gravity calculation in areas with negative ellipsoidal heights

2.2 SCALAR VERSUS VECTOR REPRESENTATION

Many interpretation procedures based on the Bouguer anomaly require validity of the Laplace differential equation (LDE). There is no problem as long as the Bouguer anomaly is formulated by a vector equation. Each component of the Bouguer anomaly vector fulfils LDE and therefore is harmonic outside the sources. However, we have rarely access to the full set of vector components. Actually, we observe the norm of a potential gradient vector, which does not meet LDE contrary to its projection onto an arbitrary but constant direction. Therefore we usually deal with scalar quantities resulting from scalar products of the gravity vector **g** and an arbitrary vector **n** as presented in Eq. (2.6).

$$\mathbf{n}(P)\cdot\mathbf{g}_S(P) = \mathbf{n}(P)\cdot\{\mathbf{g}(P) - \gamma(P) - \mathbf{g}_M(P)\} \tag{2.6}$$

Whether the Bouguer anomaly is a harmonic function or not depends on the choice of **n**. Several options exist as shown in Table 2.1, for example.

Table 2.1 Some Options for Presenting the Bouguer Gravity as a Scalar Quantity

$\mathbf{n}(P) = -\mathbf{r}(P)$	Negative radius vector	$BA(P)$ harmonic	1
$\mathbf{n}(P) = \dfrac{-\mathbf{r}(P)}{\lvert\mathbf{r}(P)\rvert}$	Unit vector along the radial direction	$BA(P)$ nonharmonic	2
$\mathbf{n}(P) = \mathbf{n}_E(P)$	Unit vector along the ellipsoid normal	$BA(P)$ nonharmonic	3
$\mathbf{n}(P) = \mathbf{n}_E = \text{const}$	Unit vector along the ellipsoid normal assumed to be constant everywhere	$BA(P)$ harmonic	4

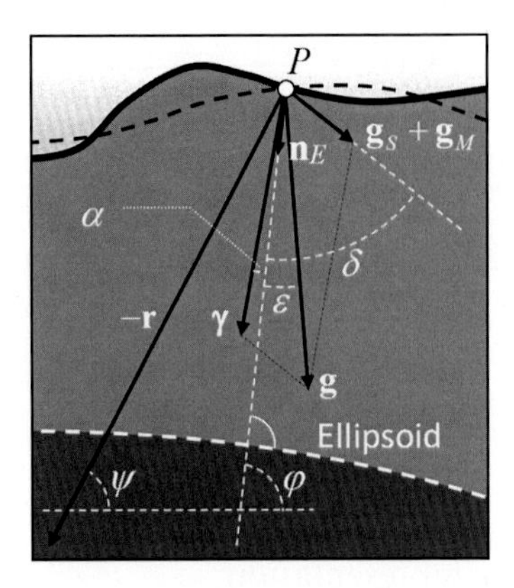

Figure 2.2 Scalar representation of the Bouguer anomaly. $\mathbf{n}_E$ *is the unit vector pointing downward along the ellipsoid normal,* α *the angle between normal gravity vector and ellipsoid normal,* ε *the angle between observed gravity vector* $\mathbf{g}$ *and ellipsoid normal,* δ *the angle between ellipsoid normal and the gravitational attraction vector of all anomalous sources inside the ellipsoid and all mass outside the ellipsoid,* $\mathbf{r}$ *the geocentric radius vector,* ψ *geocentric latitude, and* φ *geographic latitude.*

Principally, in all transformations we need to know the direction of the gravity vector **g** and not only its norm g as derived from gravity measurements. We can overcome this problem by choosing option 3 leading to a scalar formulation of the Bouguer anomaly (Fig. 2.2):

$$BA(P) = g_S(P)\cos\delta_S(P) = g(P)\cos\varepsilon(P) - \gamma(P)\cos\alpha(P) - g_M(P)\cos\delta_M(P) \tag{2.7}$$

Option 3 is common practice. The ellipsoid normal also defines the orientation of the vertical axis in the Cartesian model space used for the mass correction. This also reduces all modeling efforts because only the vertical component is required for calculating the model response. However the direction of the ellipsoid normal varies from place to place, and therefore the Bouguer anomaly derived from Eq. (2.7) does not strictly fulfill LDE. Option 4 tries to solve the problem by assuming that the ellipsoid normal points to the same direction everywhere in the survey area. This assumption is equivalent to rotating the local coordinate system at each point P such that all vertical axes are aligned, i.e., it neglects the curvature of the Earth's surface. After aligning the vertical axes, the Bouguer anomaly can be regarded as a harmonic function in planar approximation. However, it is important to keep in mind that this approximation causes errors which have to be considered in field continuation, e.g., for high accurate 3D interpolation as well as in 3D modeling. Moreover, this concept fails for

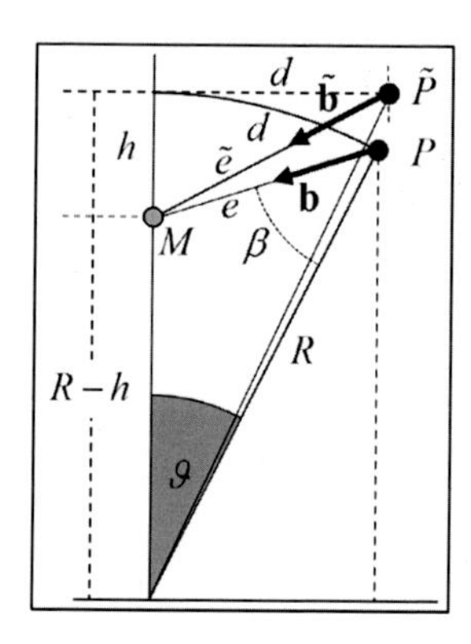

Figure 2.3 Effect of bending the Earth's surface into a horizontal plane.

large-scale investigations. Modern 3D modeling codes (e.g., Götze, 1984, http://potentialgs.com/) are able to get rid of the problem by taking the curvature into account.

We can roughly estimate the magnitude of the error. Option 4 is equivalent to bending the Earth's surface into a horizontal plane by shifting the observation point P to $\tilde{P}$ (see Fig. 2.3). Let $\mathbf{b}$ and $\tilde{\mathbf{b}}$ be the gravitational attraction vector at the observation point P and $\tilde{P}$, respectively, caused by a point source with mass M located in the depth h below the Earth's surface. According to Eq. (2.7) the Bouguer anomaly BA at a point P then reads in spherical approximation as

$$
\begin{aligned}
BA(P) &= \frac{GM}{e^2}\cos\beta = \frac{GM}{e^2}\frac{1}{e}\sqrt{e^2-(R-h)^2\sin^2\vartheta} \\
BA(\tilde{P}) &= \frac{GM}{\tilde{e}^2}\frac{h}{\tilde{e}}
\end{aligned}
\tag{2.8}
$$

with

$$
\begin{aligned}
&e = \sqrt{R^2+(R-h)^2-2R(R-h)\cos\vartheta} \quad d = R\vartheta \\
&\tilde{e} = \sqrt{R^2\vartheta^2+h^2} \\
&\sin\beta = \frac{R-h}{e}\sin\vartheta \quad \cos\beta = \frac{1}{e}\sqrt{e^2-(R-h)^2\sin^2\vartheta}
\end{aligned}
$$

R is the Earth's radius and ϑ the angle between the position vectors of M and P. The difference between the Bouguer anomalies at P and

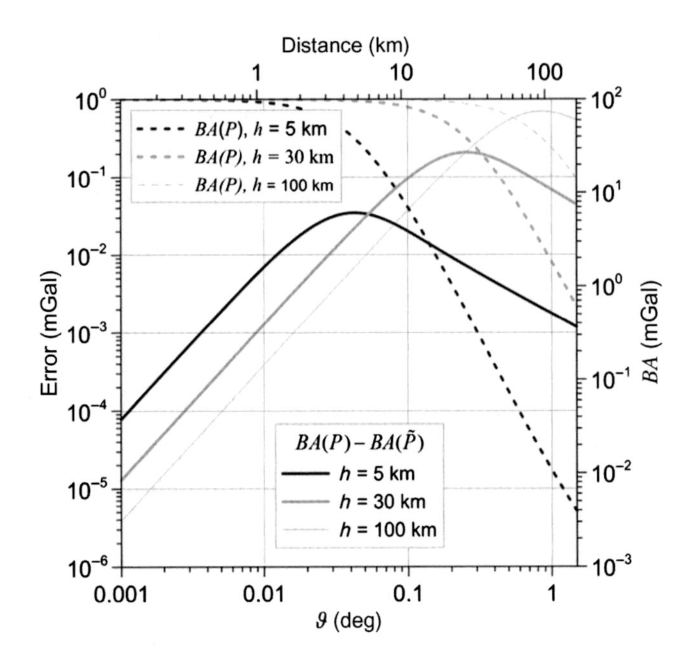

Figure 2.4 Error caused by neglecting the curvature of the Earth's surface (refer to Fig. 2.3 and Eq. 2.8).

$\tilde{P}$ depends on ϑ and h. For a mass causing an anomaly of 100 mGal (1 mGal $= 10^{-5}$ m/s^2) at a surface point with $\vartheta = 0$ the error can grow up to 1 mGal or 1% in this simple case (see Fig. 2.4).

Inherently, options 3 and 4 ignore the misalignment between the ellipsoid normal and the direction of the observed gravity vector. The error introduced by neglecting the misalignment of **g** and γ can be estimated easily. Close to the Earth's surface, the angle α between the true direction of the normal gravity vector at P and the ellipsoid normal passing P is very small. Ignoring α we get (see Fig. 2.2)

$$g = |\mathbf{g}| = \sqrt{g_{s,1}^2 + g_{s,2}^2 + (\gamma_3 + g_{s,3})^2} = (\gamma_3 + g_{s,3})\sqrt{\underbrace{\frac{g_{s,1}^2 + g_{s,2}^2}{(\gamma_3 + g_{s,3})^2}}_{\mathrm{tg}^2\,\varepsilon \ll 1} + 1}$$

$$\Rightarrow \quad g_{s,3} \cong g - \gamma$$

where the indices reflect the components of the gravitational attraction vector $\mathbf{g}_s$ of the anomalous sources in the local Cartesian coordinate

Table 2.2 Error Caused by Ignoring the Misalignment Between g and γ

ε (″)	Error (nms^{-2})	
10	11	Flat terrain
30	104	Mountainous area
60	415	Maximum estimate

system. ε corresponds to the deflection from the vertical in good approximation. Again, $g_{s,3}$ fulfills LDE in contrast to $|\mathbf{g}_s|$, that means, $BA(P)$ is harmonic. Table 2.2 represents error estimates for typical terrain scenarios.

2.3 THE GEOPHYSICAL INDIRECT EFFECT

For the first time, Meurers and Ruess (2009) calculated the Bouguer anomaly for Austria based on ellipsoidal heights derived from a precise geoid model (Pail et al., 2008), assuming a crustal density of 2670 kg/m^3 both for topography and the space between geoid and ellipsoid. The GIE is the difference between this product and the Bouguer anomaly based on orthometric heights. The GIE estimate for the Eastern Alps shows that ignoring the GIE induces an offset and, in addition, height dependent error components (Fig. 2.5).

Due to the dominating long-wavelength character of the geoid undulation the observed GIE–geoid admittance reflects on average the Bouguer gradient of -0.196 mGal/m in agreement with the estimate, e.g., by Hinze et al. (2005). In addition, Fig. 2.5C and D reveals small-amplitude signatures of the GIE, which are related to the local topography imaged in the high frequency part of the geoid undulation. Due to the latter, the GIE can vary locally by 0.1 mGal with a topographic height difference of 1 km, which is negligible in most studies. Therefore for small-scale surveys the GIE is less important. However, estimates of the isostatic state of a region will suffer from the offset directly influencing the average of isostatic anomalies.

Contrarily, the GIE is critical for large-scale investigations. Fig. 2.6 shows the expected effects for Europe based on the EGG2008 quasigeoid (Denker, 2009, http://www.isgeoid.polimi.it/Geoid/Europe/europe2008_g.html).

Ignoring the GIE will deteriorate 3D interpretation of deep sources like the lithosphere–asthenosphere boundary (LAB). Assuming a

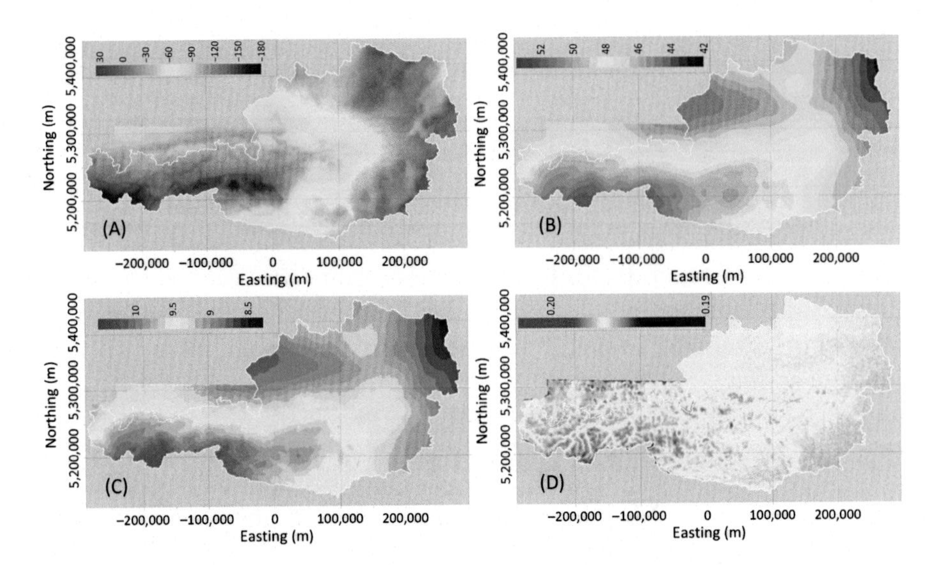

Figure 2.5 Bouguer anomaly, geoid, GIE, and GIE–geoid admittance of Austria. (A) Bouguer anomaly (mGal) based on ellipsoidal heights (Meurers and Ruess, 2009), (B) geoid (m) (Pail et al., 2008), (C) geophysical indirect effect (mGal) assuming a crustal density of 2670 kg/m^3, (D) GIE–geoid admittance (mGal/m). Gauss–Krüger coordinate system centered at 13°20′ E.

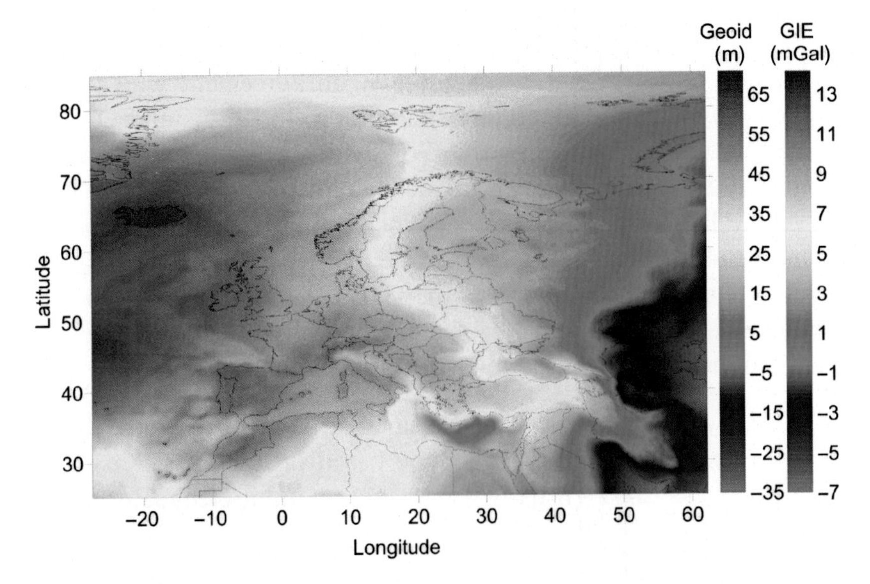

Figure 2.6 EGG2008 quasigeoid (http://www.isgeoid.polimi.it/Geoid/Europe/europe2008_g.html, Denker, 2009) and its corresponding GIE estimate in Europe based on an average GIE–geoid admittance of 0.196 mGal/m.

density contrast $\delta\rho_{LAB}$ of 50 kg/m^3 at the LAB, the GIE can be expressed as depth variation of the LAB interface:

$$GIE = \delta g_{LAB} \cong 2\pi G \delta\rho_{LAB} \delta h_{LAB} \Rightarrow$$
$$\delta h_{LAB}[\mathrm{km}] \cong \frac{GIE[\mathrm{mGal}]}{2} \text{ with } \delta\rho_{LAB} = 50\ \mathrm{kg/m^3} \tag{2.9}$$

The corresponding total LAB interface error would be 10 km approximately.

2.4 TRUNCATION ERROR OF THE MASS CORRECTION

The mass correction term in Eq. (2.2c) implicitly considers both solid rocks within topography and water (lakes, ocean, and ice) with different densities. In principle, the mass correction has to be performed over the entire Earth surface. Just by convention, it is truncated beyond the outer limit of the Hayford zone O_2 corresponding to an angular distance of roughly 1.5 degrees or 167 km. Because Eq. (2.2c) requires removing all topographic masses globally, this violates the synthetic concept and may induce a bias as well as small height-dependent errors.

Fig. 2.7 shows the effect of extending the mass correction boundary from 167 to 217 km for the Alpine area in Western Austria. Besides a bias, the differences are clearly correlated with the station elevations.

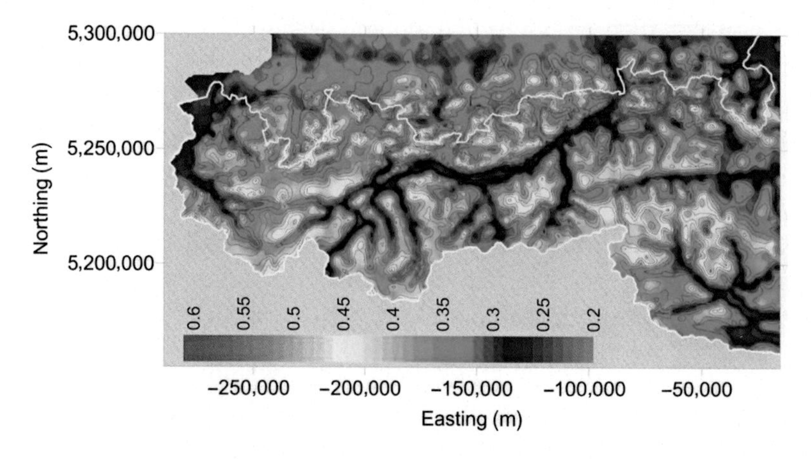

Figure 2.7 Effect of extending the mass correction boundary from 167 to 217 km for the Alpine area in Western Austria. Colors indicate the difference in mGal. Contour lines show the station elevation in 250 m intervals. Gauss–Krüger coordinate system centered at 13°20′E.

Mikuška et al. (2006a,b, 2008) studied the truncation effect in detail. They estimated the gravitational effect of both the distant terrain masses and the bathymetric relief located globally beyond the Hayford zone O_2 in spherical approximation and called it "distant relief effect" (DRE).

The distant masses turned out to produce large gravity effects globally ranging between -60 and -210 mGal with horizontal gradients up to 0.03 mGal/km and vertical gradients up to 3 mGal/km. Mikuška et al. (2006a) recommended considering the effect of distant masses for large-scale investigations or even for detailed surveys given the station heights vary remarkably. In the Alpine region, the gravity effect of the distant masses is well described by an E-W trend of 0.005 mGal/km and a bias of 105 mGal (Mikuška et al., 2006b).

In the sense of Eq. (2.2c) the real earth can be regarded (e.g., Meurers and Vajda, 2006) as synthetically composed of

- the mass of the reference ellipsoid,
- topographic surplus mass (outside the reference ellipsoid),
- water (ice) (inside or outside the reference ellipsoid),
- topographic deficit mass (inside the reference ellipsoid), and
- mass anomalies inside the reference ellipsoid.

The gravity of the reference ellipsoid is just the normal gravity used in Eq. (2.2c). The total mass of the real earth must be equal to the mass of the reference ellipsoid, because the solid spherical harmonic expansions of their gravitational potentials have identical zero-degree terms. Therefore the topographic surplus mass, the deficit mass, and all water must be balanced by the density anomalies inside the reference ellipsoid. Consequently, the Bouguer anomaly defined in Eq. (2.2c) does not only contain the signal of local sources but also the signal of all compensating mass, i.e., the 3D model space has to cover the entire Earth. The limitation of the 3D model space to smaller scales is possible as long as the gravity effect of all distant sources is smooth within the survey area and thus can be removed by trend field separation.

From this point of view, it is questionable, whether the DRE should be corrected for or not, and the issue is still under debate. The gravity effect of solid topographic mass and ocean water is isostatically compensated to a high extent by the crust-mantle boundary undulation (root and antiroot). It is reasonable to expect that the compensating masses produce similar horizontal and vertical gradients as the DRE does. If the DRE is removed then an isostatic correction should be applied as well as done in crustal balancing investigations based on isostatic anomalies. This idea has been proposed already by Hinze et al. (2005). However, crustal as well as large subcrustal mantle density variations take part in the isostatic compensation (Kaban et al., 2004). The amplitude of isostatic anomalies based on ideal isostatic models like Airy or Vening-Meinesz is significantly reduced when using real lithosphere data (Kaban et al., 1999). Therefore even the classical isostatic correction would not remove totally the gravity effect of all the mass inhomogeneity down to the upper mantle taking part in topography compensation.

It always depends on the problem we are faced to, which distant sources have to be corrected for. As mentioned earlier, in most cases 3D modeling requires a priori trend separation for keeping the 3D model space as small as possible. The horizontal DRE gradient varies very smoothly, and therefore the trend removed by the applied filter procedures includes the DRE trend too. Truncation is justified as long as it does not cause height-dependent errors. Therefore the vertical DRE gradient is more critical especially in rugged terrain because in these regions the DRE induces station height-dependent terms that cannot be eliminated by smooth trend functions. In the area of the Alps the DRE vertical gradient is about 0.5 mGal/km (Mikuška et al., 2006a,b). Elevation differences in the close surrounding of valleys amounts up to 2 km producing an anomaly of 1 mGal, which is closely correlated with topography and affects sedimentary filling estimates systematically.

2.5 NORMAL GRAVITY CALCULATION IN AREAS WITH NEGATIVE ELLIPSOIDAL HEIGHTS

Orthometric or normal heights are positive almost everywhere on the Earth surface. This allows applying closed expressions (Hofmann-Wellenhof and Moritz, 2005; Li and Götze, 2001) or series expansions (Wenzel, 1985) for the normal gravity calculation almost worldwide. This does not hold true for wide areas if ellipsoidal heights are used. Ellipsoidal heights are negative not only in some oceanic regions, but also onshore along the coast (Fig. 2.8).

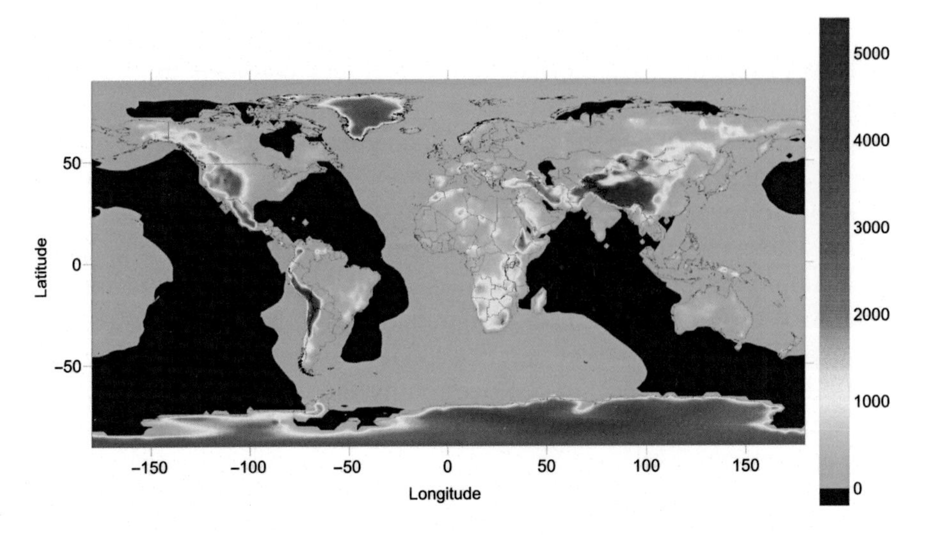

Figure 2.8 Global ellipsoidal heights (m) derived from GTOPO30 (http://www.temis.nl/data/topo/dem2grid.html) and the EGM2008 geoid undulations (Pavlis et al., 2012). Areas below the reference ellipsoid are displayed in blue.

Because the normal potential is uniquely defined by four Stokes' constants (e.g., Hofmann-Wellenhof and Moritz, 2005), no assumption is required regarding the internal mass distribution of the reference ellipsoid. Due to the nonuniqueness principle of potential theory, infinitely many internal mass distributions are able to produce the normal potential. The expansion of the normal potential into solid spherical harmonics converges down to a sphere closely surrounding the focal points of the reference ellipsoid (Moritz, 1980). According to the equivalent source principle a surface density distribution spread over the sphere of convergence exists that causes the normal potential. Thus the normal potential is harmonic far below the surface of the reference ellipsoid. However, this concept valid in physical geodesy is not suitable in geophysics because the link to the model space is missing. Normal potential is assumed to be caused by a suitable density stratification (spheroidal, ellipsoidal) within the reference ellipsoid, which serves as reference density in 3D modeling. Within the ellipsoid Poisson's differential equation holds rather than LDE because the disturbing potential is not harmonic below the reference ellipsoid surface. Taylor series approximation of normal gravity or closed expressions are no longer applicable.

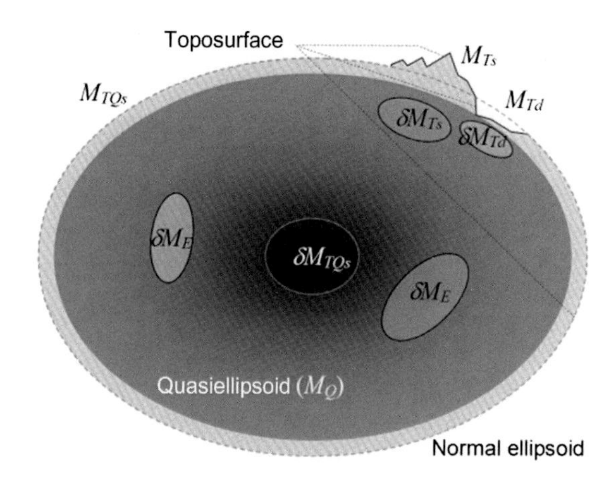

Figure 2.9 Quasiellipsoid concept (Vajda et al., 2004; Meurers and Vajda, 2006). For explanation, see text.

Therefore the common way of normal gravity determination is not possible or induces systematic errors. We need alternate reference Earth models. Vajda et al. (2004) suggest referring to a so-called quasiellipsoid. Alternatively, we can make use of ellipsoidal models presenting normal gravity at their surface (Karcol, 2011), see also Chapter 4, Normal Earth Gravity Field Versus Gravity Effect of Layered Ellipsoidal Model.

The quasiellipsoid (Fig. 2.9) is defined as a spheroid located inside the reference ellipsoid with constant distance between each surface point and its projection onto the reference ellipsoid along the reference ellipsoidal normal. Each point of the real toposurface is either on or outside the quasiellipsoid (Vajda et al., 2004). The internal mass density distribution is such that its external gravity field equals exactly to that of the reference ellipsoid. Outside the reference ellipsoid the potential field is uniquely determined by the normal gravity on the reference ellipsoid. Within the space between the reference ellipsoid and the quasiellipsoid LDE holds allowing for analytical continuation of the external potential down to the surface of the quasiellipsoid. Hence, everywhere outside and on the quasiellipsoid, the closed expressions or series expansions for the normal gravity can be applied, i.e., the components of the Bouguer gravity disturbance vector can now be calculated without an a priori knowledge of the internal density structure even at points with negative ellipsoidal heights.

The quasiellipsoid has the same total mass as the reference ellipsoid. Therefore additional anomalous mass δM_{TQs} exists in its interior that balances the mass M_{TQs} between reference and quasiellipsoid as well as the topographic surplus M_{Ts} and deficit M_{Td} mass. However, this leads to a small global- and height-dependent gravity signal (Meurers and Vajda, 2006). The Bouguer anomalies differ in both concepts by one constant- and one height-dependent term which can be estimated in spherical approximation as:

$$\delta g \approx -4\pi G\rho h' \left(1 - 2\frac{h}{R_0}\right) \tag{2.10}$$

where h' denotes the radius difference between ellipsoid and quasiellipsoid while R_0 is the radius of the reference ellipsoid. Assuming a crustal density of 2670 kg/m^3 and $h' = 0.1$ km the height dependence is estimated to be 70 nm/s^2 per km.

2.6 CONCLUSION

Only the Bouguer anomaly based on ellipsoidal heights corresponds exactly to the gravity effect of all masses below the toposurface that differ from the density of the reference Earth (if located within the reference ellipsoid) and constant density, respectively (if located outside the reference ellipsoid) at the observation point. At local and regional scales the Bouguer gravity can be regarded to be harmonic in planar approximation everywhere above and on the toposurface. The scalar representation is justified but can be associated with considerable errors at larger scales. The GIE may interfere with signals of deep (upper mantle) sources and therefore ellipsoidal heights should be used for calculating the Bouguer anomaly in large-scale studies. The truncation of the mass correction area is justified only for local- to regional-scale investigations. However the DRE is important on large scales and even sometimes at regional scales at specific locations. If applied a global isostatic correction should be considered as well to account for subsurface masses compensating the topographic load. The normal gravity cannot be calculated by applying the classical concepts in areas with negative ellipsoidal heights. We need either the quasiellipsoid approach or a global Earth model with spheroidal density stratification. In the quasiellipsoid approach the density distribution of the reference ellipsoid does not need to be known. However a global- and height-dependent signal is left in the Bouguer anomaly.

REFERENCES

Denker, H., 2009. The European Gravimetric Quasigeoid EGG2008. Eos Trans. AGU, 90(22), Jt. Assem. Suppl., Abstract CG73A-03.

Götze, H.J., 1984. Über den Einsatz interaktiver Computergraphik im Rahmen 3-dimensionaler Interpretationstechniken in Gravimetrie und Magnetik. Habilitationsschrift. Technische Universität Clausthal, Clausthal.

Hackney, R.I., Featherstone, W.E., 2003. Geodetic versus geophysical perspectives of the 'gravity anomaly'. Geophys. J. Int. 154, 35–43.

Hinze, W.J., Aiken, C., Brozena, J., Coakley, B., Dater, D., Flanagan, G., et al., 2005. New standards for reducing gravity data: the North American gravity database. Geophysics 70 (4), J25–J32.

Hofmann-Wellenhof, B., Moritz, H., 2005. Physical Geodesy. Springer, Vienna and New York, NY, 403 p.

Kaban, M.K., Schwintzer, P., Tikhotsky, S.A., 1999. Global isostatic gravity model of the Earth. Geophys. J. Int. 136, 519–536.

Kaban, M.K., Schwintzer, P., Reigber, Ch, 2004. A new isostatic model of the lithosphere and gravity field. J. Geodesy 78, 368–385. Available from: http://dx.doi.org/10.1007/s00190-004-0401-6.

Karcol, R., 2011. Gravitational attraction and potential of spherical shell with radially dependent density. Stud. Geophys. Geod. 55 (1), 21–34.

LaFehr, T.R., 1991. Standardization in gravity reduction. Geophysics 56, 1170–1178.

Li, X., Götze, H.J., 2001. Ellipsoid, geoid, gravity, geodesy, and geophysics. Geophysics 66, 1660–1668.

Meurers, B., 1992. Untersuchungen zur Bestimmung und Analyse des Schwerefeldes im Hochgebirge am Beispiel der Ostalpen. Österr. Beitr. Met. Geoph. 6, 146 S.

Meurers, B., Ruess, D., 2009. A new Bouguer gravity map of Austria. Aust. J. Earth Sci. 102, 62–70.

Meurers, B., Vajda, P., 2006. Some aspects of Bouguer gravity determination - revisited. In: Contributions to Geophysics and Geodesy, Vol. 36, Special Issue: 2nd Workshop on International Gravity Field Research, Smolenice 2006, 99–112.

Mikuška, J., Pašteka, R., Marušiak, I., 2006a. Estimation of distant relief effect in gravimetry. Geophysics 71, J59–J69.

Mikuška, J., Pašteka, R., Marušiak, I., Bielik, M., Hajach, M., 2006b. Distant relief effect and its possible impact on large-scale gravity interpretations. In: EAGE 68th Conference & Exhibition—Vienna, Austria, 12–15 June 2006, extended abstract H017, Available from: http://dx.doi.org/10.3997/2214-4609.201402092.

Mikuška, J., Pašteka, R., Mrlina, J., Marušiak, I., 2008. Gravitational effect of distant Earth relief within the territory of former Czechoslovakia. Stud. Geophys. Geod. 52, 381–396.

Moritz, H., 1980. Advanced Physical Geodesy. Herbert Wichmann Verlag, Karlsruhe.

Moritz, H., 1990. The Figure of the Earth. Herbert Wichmann Verlag, Karlsruhe.

Naudy, H., Neumann, R., 1965. Sur la definition de l'anomalie de Bouguer et ses consequences practique, Geophys. Prospect., 13. pp. 1–11.

Pail, R., Kühtreiber, N., Wiesenhofer, B., Hofmann-Wellenhof, B., Of, G., Steinbach, O., et al., 2008. The Austrian Geoid 2007. Vgi - Österreichische Zeitschrift für Vermessung und Geoinformation 96 (1), 3–14, Österreichische Gesellschaft für Vermessung und Geoinformation (OVG), ISSN 0029-9650.

Pavlis, N.K., Holmes, S.A., Kenyon, S.C., Factor, J.K., 2012. The development and evaluation of the Earth Gravitational Model 2008 (EGM2008). J. Geophys. Res. 117, B04406. Available from: http://dx.doi.org/10.1029/2011JB008916.

Tsuboi, C., 1965. Calculations of Bouguer anomalies with due regard the anomaly in the vertical gradient. Proc. Jap. Acad. Sci. 41, 386–391.

Vajda, P., Vaníček, P., Meurers, B., 2006. A new physical foundation for anomalous gravity. Stud. Geophys. Geodesy 50 (2), 189–216. Available from: http://dx.doi.org/10.1007/s11200-006-0012-1.

Vajda, P., Vaníček, P., Novák, P., Meurers, B., 2004. On evaluation of Newton integrals in geodetic coordinates: exact formulation and spherical approximation. Contrib. Geophys. Geodesy 34, 289–314.

Vogel, A., 1982. Synthesis instead of reductions—new approaches to gravity interpretations, Earth Evolution Sciences, 2. Vieweg, Braunschweig, pp. 117–120.

Wenzel, F., 1985. Hochauflösende Kugelfunktionsmodelle für das Gravitationspotential der Erde. Wiss. Arb. Fachr. Vermessungswesen Univ, Hannover, p. 137.

CHAPTER 3

Some Remarks on the Early History of the Bouguer Anomaly

Ján Mikuška[1], Roman Pašteka[2], Pavol Zahorec[3], Juraj Papčo[4], Ivan Marušiak[1] and Martin Krajňák[2]

[1]G-trend, s.r.o., Bratislava, Slovak Republic [2]Comenius University, Bratislava, Slovak Republic [3]Slovak Academy of Sciences, Banská Bystrica, Slovak Republic [4]Slovak University of Technology, Bratislava, Slovak Republic

3.1 INTRODUCTION

We think that the present-day theory and practice of ground gravity method can be seen in the proper light only when we know as much as possible about its basic elements, namely about the Bouguer anomaly, which we consider *a most important notion*. We think we know how to calculate this quantity, but do we know enough about its historical development and background?

Unfortunately, this topic is rather large and therefore an ambition to draft out the history of such a subject matter, which would be both complete and the same time concise, would be unrealistic. For that reason, instead of providing an overall review, we would like to present some pieces of the mosaic, derived from the earlier contributions of various authors and perceived by us as being important. In both its content and form, this retrospective is a consequence of many discussions among its authors over the years. Needless to say those discussions were always followed by searching for the literature and then searching in the literature, establishing a seemingly never-ending process.

We realize that writing about history should not be confused for writing a critique, yet it will become evident later that some critical approach will have still to be involved. Presumably, one can hardly analyze the development of a subject without touching intimately the subject as such.

Understanding the Bouguer Anomaly. DOI: http://dx.doi.org/10.1016/B978-0-12-812913-5.00002-6

We appreciate that today the majority of the exploration geophysicists understand the Bouguer anomaly simply as "the difference between the observed gravity and the modeled or predicted value of gravity at the station" (Hinze et al., 2005, p. J28, to quote here just the newest of many important papers advocating the concept of station anomaly). In fact here the authors use the term "gravity anomaly" while the terms "Bouguer gravity anomaly" or "Bouguer anomaly" occur elsewhere in their text. At the first glance it is obvious that such a definition is rather vague, but something like this we will have to face throughout all the following parts of our contribution. Moreover, some vagueness or ambiguity will later silently emerge from our retrospective possibly as one of the perceived characteristic features of the Bouguer anomaly.

To keep our text focused on the history (and especially on the earlier history) of the topic we will have to ignore some significant issues as are the influence of the Earth's atmosphere (Ecker and Mittermayer, 1969, and others) or the so-called geophysical indirect effect (Chapman and Bodine, 1979, and others). As well, we will not touch in a greater detail the concept of isostasy (e.g., Watts, 2001) which we rather consider an independent matter. Further, we will not discuss airborne or satellite gravimetry at all, and we will mention the underground and underwater conditions only marginally. We will focus almost exclusively on the ground (surface) measurements of gravity and the Bouguer anomaly calculated on their basis.

3.2 THE EARLY DAYS: GEODESIC MISSION TO ECUADOR (THEN PERU) AND THE BOOK OF BOUGUER (1749)

It is well known that this famous mission commenced in 1735. Three French academics, namely Pierre Bouguer, Charles Marie de La Condamine, and Louis Godin; two Spanish naval officers, Antonio de Ulloa y de la Torre-Giral and Jorge Juan y Santacilia, were the five prominent mission members. On the other hand, the conclusion of the mission cannot be clearly specified because of serious quarrels among its members. For instance Bouguer returned home in 1744. The primary aim of the mission was "measuring an arc of the meridian near the equator in order to compare the corresponding length of a degree with that which had been obtained from the French arc by Jean Picard and by Jacques Cassini" (Todhunter, 1873a, p. 93). They also measured the Earth's gravity at different elevations above sea level by a one-second pendulum.

The most famous and presumably the most important report about the mission results is given in the book of Bouguer (1749). Before we proceed further with our own discussion we would like to give here two citations. The first one is Todhunter's (1873a, p. 248) comment which we quote since we consider his description of the Bouguer's book illustrative: "Bouguer treats on the diminution of attraction at different heights above the level of the sea. He finds that on a mountain at the height h above the level of the sea, the attraction is proportional to

$$(r - 2h)\Delta + \frac{3}{2}h\delta \tag{3.1}$$

where r is the Earth radius, Δ the Earth mean density, and δ the density of the mountain. This is the first appearance of the formula, which has now passed into elementary books." We can only agree with Todhunter that the importance of the expression (3.1) is indeed extraordinary, especially from the aspect of our topic. The corresponding part of the Bouguer's original text is reproduced in Fig. 3.1.

The second citation is a statement of Bullen (1975, pp. 13, 14): "… the second and third terms" in one of Bullen's equations very similar to the expression (3.1) "are associated with what are now, in the reduction of gravity observations, called the free-air and Bouguer corrections, respectively." This statement of Bullen sheds light on the Bouguer anomaly definition as we know it today and on its direct relation to Bouguer (1749). On the other hand, from the aspect of terminology, it is interesting to note that Bullen associates the expression (3.1) with the "reduction" of gravity while he calls its terms Nos. 2 and 3 "corrections."

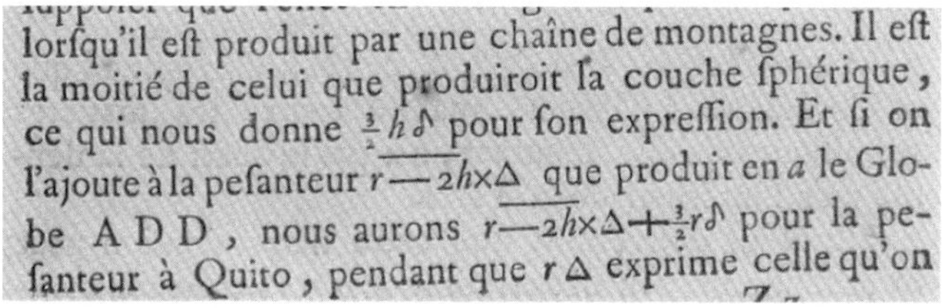

lorſqu'il eſt produit par une chaîne de montagnes. Il eſt la moitié de celui que produiroit ſa couche ſphérique, ce qui nous donne $\frac{3}{2}h\delta$ pour ſon expreſſion. Et ſi on l'ajoute à la peſanteur $\overline{r-2h}\times\Delta$ que produit en *a* le Globe A D D, nous aurons $\overline{r-2h}\times\Delta+\frac{3}{2}r\delta$ pour la peſanteur à Quito, pendant que $r\Delta$ exprime celle qu'on

Figure 3.1 The original text with the original expression (3.1) (Bouguer, 1749, p. 361). Please note that horizontal line was used in that time instead of parentheses. Also note that there is a misprint in the resulting formula (there should be h instead of r in the last term). Todhunter (1873a) quotes the expression correctly and does not mention the misprint at all.

The second and third terms can be better seen in expression (3.1a) where we removed the parentheses from Eq. (3.1), just to make it demonstrative:

$$r\Delta - 2h\Delta + \frac{3}{2}h\delta \tag{3.1a}$$

Let us now examine the expression (3.1a) in a greater detail, with some help from Bouguer's own text. First let us multiply all three terms by the factor $\frac{4}{3}\pi\gamma$, where the last symbol stands for the gravitational constant with the known value of 6.674×10^{-11} kg/m^{3}per s^2 (Petit and Luzum, 2010, p. 18). We get

$$\frac{4}{3}\pi\gamma r\Delta - \frac{8}{3}\pi\gamma h\Delta + 2\pi\gamma h\delta \tag{3.1b}$$

It is now evident that the first term of Eq. (3.1b) represents the gravitational effect of a homogenous sphere with radius r and density Δ, calculated at its surface. Taking the γ value as quoted earlier, Δ equal to 5515 kg/m^3 (Cox, 2002, p. 12) and $r = 6371000$ m (the mean radius of the oblate ellipsoidal reference figure, Cox, 2002, p. 240) we obtain for the Bouguer's first term the value about 981938 mGal (1 mGal $= 10^{-5}$ m/s^2). Just as a matter of interest, the modern GRS80 gravity formula (Moritz, 1988, p. 353) results in the value of 981938 mGal for the latitude of approximately 61 degrees. In other words, even if we disregard the Earth rotation and if we use the spherical approximation of the Earth, with constant density, which is rather rough, the estimated gravitational effect falls well within the acceptable limits.

Further we would like to focus on the origin of the second term in Eq. (3.1) or (3.1a). Among others its meaning was correctly understood by Bullen (1975, pp. 13, 14), but the question is how had Pierre Bouguer arrived at it? Bouguer (1749, p. 358) wrote, when commenting upon the shortening of the pendulum length by 1/1331 between an island in the river Inca situated at a low elevation, and the city of Quito, elevated above the lower station approximately by 1/2237 of the Earth radius, that the pendulum shortening was not too far from reciprocal proportionality to the square of the height difference "since we know that squares of quantities which differ only slightly one from another will change twice in proportion to those quantities." We consider this statement as quite essential from the aspect of proper understanding Pierre Bouguer's thoughts. What did he mean? We presume that Bouguer spoke here about the differential $dy = 2xdx$ of the function $y = x^2$ since that notion as well as the mathematical tool had been then already known, thanks to Leibniz and Newton. Thus, if $x = 1$ and $dx = 1/2237$, for the differential in question we will have

$$dx^2 = 2 \times 1 \times \frac{1}{2237} = \frac{2}{2237} = \frac{1}{1118.5}$$

provided that the decrement of the pendulum length would be proportional solely to the increase of the distance from the Earth center. Then his pendulum should have been shorter at the Plaza Grande in Quito rather by 1/1118.5 than the determined rate of 1/1331. Nevertheless, in his evaluation of the expected gravity at Quito he decided to consider the inverse-square principle in the form of the second term in his expression (3.1), and later to introduce the corrective third term. This we regard the key step in Bouguer's thinking. In fact Bouguer did not write explicitly how he got the actual form of the second term but he most likely calculated the ratio

$$\frac{g_Q}{g_I} = \frac{r^2}{r^2 + 2rh + h^2}$$

where g_Q and g_I stand for the quantities proportional to gravity at the stations in Quito and the river Inca island, respectively, and h means the elevation difference between those two stations.

Then, supposing that g_I is proportional to $r\Delta$, he obtained

$$g_Q - g_I = \Delta \frac{-2h - \frac{h^2}{r}}{1 + \frac{2h}{r} + \frac{h^2}{r^2}} \approx -2h\Delta \tag{3.2}$$

after neglecting the fractions with r or r^2 since $h \ll r$ (Bouguer, 1749, p. 360, wrote that h in proportion to r was "très-petite"). Instead of gravity g we could have used pendulum length l to which g is proportional and which was actually measured. Well, the mentioned simplification, especially regarding the term $-h^2/r$ in the nominator, does not seem to be legitimate but, on the other hand, the difference between the approximation $-2h$ and the full fraction in Eq. (3.2) (not considering the multiplication by Δ!) represents only 0.067%, 0.112%, and 0.208% for the elevations of Quito, Pichincha, and Mount Everest, respectively.

According to Bouguer (1749, p. 360) the second term of our expression (3.1a) describes the diminution of gravity when we move from the river Inca island to Quito only if the Earth "would end" at the level of the lower station. All this should change however, if we add a new spherical layer to the Earth with the thickness h. As a result the new (hypothetical) gravity at Quito would have been equal to $(r + h)\Delta$ (provided that the density of the added spherical layer would remain the same), instead of $(r-2h)\Delta$ which we should have measured if there were no rock-filled spherical layer beneath our feet at Quito. The effect of the (complete) spherical layer calculated at Quito would then be $3h\Delta$ provided that the density *was* the same as for the whole Earth. But Bouguer noticed this was not the case and therefore the final form of the spherical layer effect he estimated as $3h\delta$, where δ stands for the density of what we now would call topographic masses. Yet he instantaneously recognized that the Cordillera could not produce an effect comparable to the one of a complete spherical layer. Trying to approximate the true topographic effect, he first considered a roof-like model with the ridge angle of 90 degrees between the roof sides which should produce about 1/4 of the complete spherical layer effect (this Bouguer states without derivation but it can be easily checked; we have found it to be essentially correct; see Fig. 3.2).

Bouguer then appreciated that in reality the base of the Cordillera must be 80–100 times greater than its height, and as a result, the roof-ridge-angle should increase from the previously considered 90 degrees

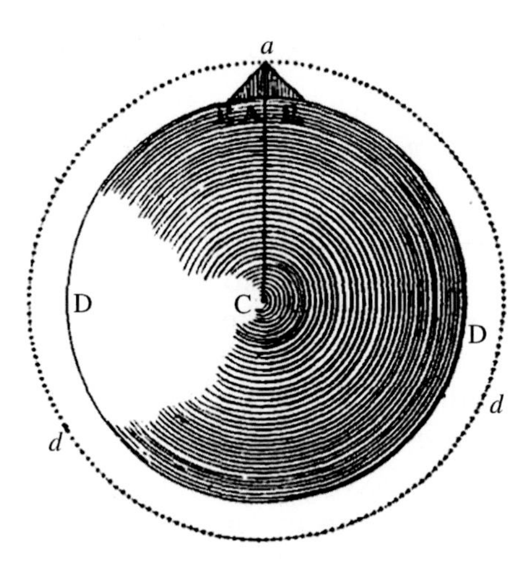

Figure 3.2 Bouguer's estimation of the Cordillera's gravitational effect at the station in Quito, by the roof model. In his first attempt, Bouguer used a roof-ridge-angle of 90 degrees as shown. Reproduced from Bouguer, P., 1749. La figure de la terre. Paris: Charles-Antoine Jombert, 394 p.

to about 170 degrees. Moreover, in reality near Quito, the roof-top model is far from being perfect. Instead the true model top should have a width of about 10–12 miles. Bouguer then concluded the following: "Therefore we can presume, without any risk of being mistaken, that the effect of the mountain belt reaches its possible maximum, i.e., the half of the effect which would be produced by the complete spherical layer, which we can express as

$$\frac{3}{2}h\delta \tag{3.3}$$

If we add the expression (3.3) to the value of $r\Delta - 2h\Delta$, which is supposed to be caused by the spherical Earth without topography, the gravity at Quito can be expressed as (3.1), while $r\Delta$ expresses the gravity encountered at sea level." Let us now sum up the Bouguer's findings.

We already have mentioned that Bouguer measured the one-second pendulum lengths in Quito (he mentioned Plaza Grande) and on a small island in the stream Inca, a tributary of the present-day Rio Blanco (the latter he then called "rivière des Émeraudes" although today that name, namely Rio Esmeraldas in Spanish, is used for the river originating from the mentioned Rio Blanco and another river, Rio Guayllabamba, some 20 km downstream). In addition, there was the third place, the summit of the Pichincha volcano on the south-eastern slopes of which the City of Quito had been established. Based on his geodetic and astronomical observations Bouguer estimated the elevations of those three places. For the river island within the Inca stream it was 40–42 toises (Bouguer, 1749, p. 166) although he originally believed that it had to be about 30 toises (Bouguer, 1749, p. 161); for Quito and Plaza Grande, he found the elevation of 1466 toises; and finally for the Pichincha summit he gave 2434 toises. With 1 toise being equal to 1.949 m, these elevations then were approximately 80, 2857, and 4744 m, respectively. It should be mentioned here that today the summit of the Pichincha volcano has a different shape resulting from eruptions, which have taken place in the meantime so that Bouguer's value of the elevation cannot be checked, while the other two above-the-sea-level elevations seem to be more or less acceptable as approximations, according to the data available today.

In fact Bouguer found that his pendulum was shorter in Quito and shortest at the summit of the Pichincha volcano when compared to its length measured at an elevation close to sea level (i.e. his Inca island station). He had immediately concluded that the force of gravity changed depending on the distance from the Earth center. "That force decreases with increasing elevation" (Bouguer, 1749, p. 357). In Quito he observed that his pendulum was shorter than it had been at the lowest station by 33/100 of the French line (1 line = 2.2558 mm) or by 1/1331. At the summit of Pichincha it shortened by an additional 19/100 of the French line so that it was shorter there by 1/845 compared to its length at the lowest station. We should only add here that his fractions 1/1331 and 1/845 were of course approximate. Bouguer readily recognized that those differences in the pendulum length (i.e., quantities proportional to gravity) cannot be ascribed to the centrifugal force, in spite of the fact that it should act in the required sense. Bouguer estimated the amplitude of its centrifugal component to be much less than the observed difference. He offered expression (3.1) as a way how to estimate the gravity decrease with increasing height of the measuring place. In our opinion, expression (3.1) represented a genuine generalization of those few measured values. Today we can interpret Eq. (3.1) also in terms of the expected or theoretical gravity estimation. Therefore we understand that with his expression (3.1) Bouguer laid the foundation of what has become known today as the Bouguer anomaly as well as the foundation of the gravity method of (applied) geophysics.

It is important to note that the second and the third terms of the expressions (3.1) or (3.1a), namely Eq. (3.2) and expression (3.3), were based on spherical approximation, as well as the first term, as shown earlier. This remains valid even if we acknowledge Bouguer's auxiliary estimations using his roof model as shown in Fig. 3.2. The terms like "the Bouguer infinite slab" or "plate" which are now associated with the third term were in fact developed later after Young (1819) and Poisson (1833) have done their work regarding the well-known expression for the gravitational effect in question, see later. Here we should add that even in the Poisson's approach we cannot see any straightforward replacement of some part of an idealized topography with a spherical base by a (horizontal and infinite) plate or slab. In fact, sometimes the things in the past happened a bit differently than they are seen today.

Immediately after having arrived at the expression (3.1), Bouguer applied it to derive an estimate of the topographic density. Taking the pendulum shortening between the Inca island and Quito 1/1331 and $h/r \approx 1/2237$ he got (Bouguer, 1749, pp. 362)

$$\frac{1}{1331} = \frac{2h\Delta - \frac{3}{2}h\delta}{r\Delta} \Rightarrow \frac{1}{1331} = \frac{1}{2237} \times \frac{2\Delta - \frac{3}{2}\delta}{\Delta} \Rightarrow \delta = \frac{850}{3993}\Delta \quad (3.4)$$

Today we know that $\Delta \approx 5515\ \text{kg/m}^3$ (Cox, 2002) and thus Eq. (3.4) estimates the topographic density as $\delta \approx 1174\ \text{kg/m}^3$, which is not in a good agreement with the possible densities of the topographic masses. Thus from the aspect of our present-day knowledge, Bouguer's estimation of the density of the topographic masses indicated that there should be some other reason for the gravity change between those two places in addition to the ones described by the second and third terms of Eq. (3.1). To sum it up, we can consider the density estimation in Eq. (3.4) as rather unsuccessful. We will discuss this in a greater detail later.

After having discussed the very foundations of our subject we will now proceed a couple of decades onward and discuss some of the most significant issues preceding the establishment of the gravity method as a part of geophysics or applied geophysics.

3.3 REDUCTIONS OF GRAVITY OR THE PENDULUM LENGTH TO THE SEA LEVEL

3.3.1 The So-Called Bouguer Reduction

It is not easy to retrace the beginnings of the historically important concept of reductions of gravity to the sea level which has had such a controversial impact on the development of the Bouguer anomaly concept. Let us begin with the discussion of allowing for the ground masses between the station and the sea level, though historically, it seems to have been preceded by the allowance for the rest of the Earth below the sea level.

For instance Lambert (1930, pp. 137, 138) wrote: "It does not appear, however, that Bouguer recommended for general use what is now known as Bouguer's method. The first use of it as a general method for reducing to sea level appears to be due to Thomas Young who said, in effect, that the free-air method had hitherto been used but that he recommended an allowance for the matter between the station and sea level which matter might to a first approximation be treated as an infinite slab. Poisson also recommended the same method. The method is sometimes and more appropriately named after Young or Poisson, but more often and less appropriately after Bouguer."

We assumed that some of the Lambert's statements quoted earlier might have been at least partly based on Helmert, although Helmert's name was not mentioned here, but, in his "Bibliographical Notes," Lambert (1930, p. 176) refers to some others of the Helmert's works. In fact, in the fundamental textbook of Helmert (1884, p. 166), we can read more or less the same as in the above citation of Lambert (1930). Regarding the terminology associated to the allowance made for the rock-mass between the station and sea level, Helmert (1884, p. 166) wrote the following: "This relation is called the Young rule as well as the expression of Poisson for flat terrain. Anyway we will call it after Bouguer who was the first to study such relations." So, we can ask—was the technical term "Bouguer reduction" first introduced by Helmert? Likely it was. And, moreover, it is interesting to note that this happened probably merely thanks to the geodetic concept of reducing gravity from the Earth surface to the sea level.

According to Helmert (1884) and Lambert (1930), Young (1819) seems to be the first one who considered the attraction of rock masses between the gravity station and the sea level. But this has not necessarily to be the case and there are at least two reasons for some doubt. First, Young's own words (Young, 1819, p. 93) "... for example, in the allowance made for the reduction of different heights to the level of the sea, which has usually been done without any consideration of the attraction of the elevated parts, interposed between the general surface and the place of observation ...." He actually wrote "has usually been done" instead of, say, "has always been done." Second, Todhunter (1873b, p. 490) wrote that "Dr. Young's rule ... coincides with the formula originally given by Bouguer and reproduced by D'Alembert ... Dr. Young does not refer to any preceding writer." It is then obvious that Young had not a good reputation regarding quoting his predecessors. As a consequence, knowing that Bouguer (1749) did not reduce his gravity measurements to the sea level, to decide whether Young (1819) was really the first to use this kind of reduction would require further investigation.

Still Young's estimations were very interesting. Young (1819, p. 93) continues: "It is however obvious, that if we raised on a sphere of earth

$$K' = 2\pi f\rho'(c + h - \sqrt{c^2 + h^2}),$$

en désignant cette force par K'. Mais, en général, l'épaisseur verticale de la couche attirante est petite, eu égard à son rayon horizontal; si donc on néglige h^2 par rapport à c^2, on aura simplement

$$K' = 2\pi f\rho' h.$$

Figure 3.3 Poisson (1833) estimating formula. From his original sketch, however, we can see that he had actually in mind rather a model bounded by two spherical surfaces than a slab. The infinite horizontal slab and its calculated attraction, were in fact Poisson's approximations to the volume depicted in his Figure 59 and to its corresponding gravitational attraction.

a mile in diameter, its attraction would be about 1/8000 of that of the whole globe, and instead of a reduction of 1/2000 in the force of gravity, we should obtain only 3/8000, or three fourths as much ...." And finally: "Supposing the mean density of the Earth 5.5, and that of the surface 2.5 only, the correction, for a tract of table land, will be reduced to $1 - \frac{3}{4} \times \frac{2.5}{5.5} = \frac{29}{44}$, or 66/100 of the whole." After recalculation based on the present-day data his fractions become: $\frac{1}{8000} \rightarrow \frac{1}{7931}$, $\frac{1}{2000} \rightarrow \frac{1}{1975}$, but $\frac{66}{100} \rightarrow 0.66056$! The last figure deserves special attention: $0.3086 - 0.04193*2.5 = 0.20385$ and $0.20385/0.3086 = 0.66056$. It is evident that in estimating the impact of the attraction of the rocks between the observation point and the sea level Young achieved a considerably high accuracy.

Poisson (1833), on the other hand, is known to have derived the formula which has since been in use and which obviously supported the "introduction" of the concept of an infinite horizontal slab, or simply Bouguer slab, into geodetic and gravimetric practice (Fig. 3.3).

We conclude that, as a matter of paradox, neither Bouguer nor Young and even Poisson primarily considered the model of a horizontal infinite plate. The impression remains that as if they all were standing on "terra firma sphaerica." Therefore it is not simple to ask who introduced the "Bouguer plate" model and when it happened—it is rather difficult to find.

3.3.2 The So-Called Free-Air or Faye Reduction

When the free-air reduction was introduced into geodetic practice? Heiskanen and Vening Meinesz (1958, p. 150) tried to explain why this reduction often points to the name of Faye and then gave some time frame. They wrote: "The free-air reduction is often called Faye's reduction after the man who called attention to it. ... This kind of reduction was frequently used in the eighteenth century and at the beginning of the nineteenth; important work of Stokes and Faye was based on it." In fact Stokes (1849, p. 673) refers to it: "... and observed gravity be reduced to the level of the sea by taking account only of the change of distance from the earth's centre." But should Faye's work be considered important from the aspect of the free-air reduction? We are not sure. From the Faye's contributions one can deduce that this kind of reduction had been in use a long time before he wrote his memoirs, and that Faye himself apparently did nothing in order to either strengthen or weaken it. On the other hand, Faye did not want the attraction of the intermediate rock being subtracted from the measured gravity. He wrote: "The continent has been compensated, almost entirely, by disturbances in thickness of the rigid crust under the continents and therefore it is not necessary to account for it ..." (Faye, 1880b, pp. 1443, 1444). Here he also referred to his earlier paper (Faye, 1880a) where he had pointed out this concept for the first time. It is really interesting to read what he wrote later (Faye, 1880b, p. 1445): "It is however necessary to realize that, if the thickness of the continents above the sea level will not be taken into consideration, it would not be the same as, for instance, the mass of the Great Pyramid in Egypt, provided that the pendulum measurement were performed at its summit. So, after the reduction of the pendulum length at the sea level according to formula (3.5)

$$l_0 = l + \frac{2hl}{R} \tag{3.5}$$

where l_0 is the pendulum length reduced at the sea level, l is the length measured at the pendulum station at the elevation h above the sea level, and R is the Earth radius; it will be necessary to subtract the attraction of the pyramid above the Earth surface. Likewise, when Bouguer brought his pendulum at the summit of the Pichincha volcano about 1500 m above the level of the terrain in Quito, he should have accounted for the attraction of this mountain on his pendulum." In his subsequent papers Faye (1883, 1895) did not seem to introduce anything different with regard to what we have discussed here. There were some other concepts of Faye, however, which we will briefly mention and comment later.

To sum up, we have not been successful in finding any clue regarding the earlier development of the free-air reduction within either the papers of Young (1819), whom we already quoted earlier and Stokes (1849) or within the memoirs of Faye (1880a,b, 1883, 1895), or elsewhere.

3.3.3 When and by Whom the Reductions to the Sea Level Were Introduced Remains at Least Partly Unknown

Was Thomas Young really the first using what is now known in literature as Bouguer reduction? And who was the first using the so-called free-air or Faye's reduction? Those questions remain unanswered at present. We know of course that the background of both reductions come conceptually from Bouguer (1749) with only one important issue to be repeated here—Bouguer did not intend using those procedures in order to reduce his Quito or Pichincha pendulum lengths to the sea level. This should be kept in mind.

3.3.4 The First Initiative Against Reductions to the Sea Level

Hayford and Bowie (1912) seem to be the first specialists who looked at the problem of reductions differently. After they calculated the theoretical value of gravity at the sea level they computed "the correction for elevation ... of the station" according to simple expression (3.6), values in dynes, elevation H in meters,

$$-0.0003086H \tag{3.6}$$

which has negative sign, likewise the Bouguer's second term in expression (3.1) (Hayford and Bowie, 1912, p. 13), but in contrast to the positive sign used, e.g., in formula (3.5) of Faye. The correction (3.6) is applied to the value calculated according to Eq. (3.7), see later. It is obvious that "... (3.6) is the reduction from sea level, to the station, a correction to the theoretical value not the observed value. It takes account of the increased distance of the station from the attracting mass, the earth, as if the station were in the air at the stated elevation and there were no topography on the earth" (Hayford and Bowie, 1912, p. 72).

Before we continue with the discussion of their next related step, i.e., their allowing for topography, we should note that Hayford and Bowie (1912) had decided to (1) correct for the topography and its isostatic compensation simultaneously, and to (2) take into consideration the whole globe up to the antipodes of the pendulum station. Having this in mind we can continue the previous quotation: "The correction for topography and compensation was computed with the new reduction tables. This is also a correction to be applied to the theoretical value at sea level." The latter statement is in fact not easy to understand since the authors write elsewhere (Hayford and Bowie, 1912, p. 28): "All tabular values are the vertical components of the attraction upon a unit mass at the station ...." The only plausible understanding seems to be that their phrase "theoretical gravity at sea level" represents the term they use for the output of the Helmert's well-known formula for γ_0 derived in 1901, namely

$$\gamma_0 = 978.046(1 + 0.005302 \sin^2 \phi - 0.000007 \sin^2 2\phi) \tag{3.7}$$

(Hayford and Bowie, 1912, p. 12; values of γ_0 are in dynes and ϕ is termed as "latitude") and that this does not mean that the correction for topography and its compensation should be applied at the sea level. Hayford and Bowie (1912, pp. 72, 73) continue: "Usually corrections are applied to the observed values of the intensity of gravity to reduce them to sea level and to correct for the supposed influence of topography. In this publication the corrections are applied to the theoretical value of the intensity of gravity at sea level to obtain the theoretical value at the station, a value which is directly comparable with the observed value. This seems to the authors to be a more logical method and more conductive to clear thinking than the usual method." In the times to come, many specialists have agreed the above-quoted approach is the only one compatible with geological interpretation of gravity data. On the other hand, there have been also many who either ignored it or even openly continued in supporting the original reductions to the sea level. We will give some brief examples later.

Bullard (1936) uses the term "reduction" or "to reduce" at least in three different meanings of which we will quote corresponding examples within Section 3.4.8. On the other hand, although he mentions the work of Hayford and Bowie (1912) quite frequently, he does not comment their refusing to reduce the measured gravity to the sea level. In fact he has not touched the problem of such reductions in the Bullard (1936) paper at all. Notwithstanding the assumption that he calculated and interpreted his anomalies as station quantities can be based on other kind of evidence. For instance, Bullard (1936, p. 501) writes: "The observed and calculated values of g at all the stations ...."

In addition, he evidently considers the uneven Earth surface in his interpretation (Bullard, 1936, p. 507 and following pages). Assuming this, however, we must consider his quantity called "the difference between gravity at sea level and at the height, h, of the station, neglecting the attraction of the topography" equal to "$+0.3086 \times 10^{-3}h$," h being in meters (Bullard, 1936, p. 501), as inconsistent since it should have been negative had he understood the problem as Hayford and Bowie (1912) did.

3.4 ADDITIONAL DISCUSSION

3.4.1 Bouguer (1749)

3.4.1.1 Some Interesting Quotations

Let's go back to Bouguer's own text. He among others wrote (Bouguer, 1749, p. 357): "The experiments with the pendulum which we performed in Quito, as well as at the summit of the Pichincha mountain, tell us that gravity is changing depending on the distance from the Earth centre."... "However, the measured differences cannot be ascribed to the centrifugal force." Perhaps this was the first experimental confirmation of decreasing gravity with elevation, which had been predicted earlier by Newton. Later Bouguer (1749, p. 358) continued: "But why our experiments all the time yield a relation which does not completely satisfy the quadratic condition?" And subsequently on the same page we can read: "We will possibly find the solution to this difficulty if we notice that the Cordillera is forming something like an 'other Earth' and, from some aspects, this must be the same as if the Earth surface was moved to the higher elevation or to the greater distance from the Earth centre."

3.4.1.2 Prediction or Reduction?

We understand that Pierre Bouguer was capable of predicting gravity at elevated stations. On the other hand, he evidently had no ambition of reducing the values measured at the tops of the topographic forms to any "datum plane" or to the sea level. Actually, by his expression (3.1) or (3.1b) he in fact offered a way how to estimate what we would now call the theoretical gravity, namely by calculating the gravitational effect of a "normal Earth" though today we would probably not accept the effect of the homogenous sphere as the first or the principal component of it (see Section 3.5.1.3). To sum it up: although we have found mentions about reducing the distances measured for determining the length of a meridian degree at sea level (Bouguer, 1749, p. 167 and following) we did not come across any mention of reducing gravity (or pendulum length) to sea level within Bouguer's book. On the other hand, Heiskanen and Vening Meinesz (1958, p. 153) saw things a bit differently: "The effect of the Bouguer's plate diminishes the effect of the free-air reduction by about one-third. This reduction was originated by Bouguer, who derived the formula in his work "La figure de la terre" in 1749. He used this reduction to compare the gravity values observed on the plateau of Quito and the neighboring seacoast of Peru." We interpret this final statement as being just the authors' assumption and actually this was not the case. We can only repeat that in fact we have found no indication, direct or indirect, that Bouguer (1749) wanted to compare his pendulum lengths reduced to the sea level.

3.4.1.3 Bouguer's First Term

Bouguer's prediction was based on estimating the gravitational effect of his "normal Earth" consisting of a sphere and the topography. His sphere had the radius of about 6391632 m, was characterized by constant density Δ and the gravitational effect at its surface was given by the first term of expression (3.1b). In contrast with the ellipsoid of revolution which we use today and which represents a surface with constant gravity potential but with different values of gravity at different latitudes (i.e., equipotential surface), the surface of the Bouguer's sphere not only had the same gravitational potential (as we would say today) but also the same gravitation at its surface. Thus Bouguer's "normal Earth" was equigravitational (neglecting the centrifugal component). We realize that the mentioned qualities of the ellipsoid and especially of the approximation sphere may seem unimportant yet we are convinced that their impact was in reality quite serious as we will see later in our discussion of Bouguer's density estimate.

3.4.1.4 Bouguer's Second Term: The Free-Air, Faye, or Height Correction/Reduction Is Actually Due to Bouguer

The second term of Eq. (3.1) or (3.1b) represents what is now known as the free-air, Faye or height correction/reduction. Therefore it should be credited to Bouguer as identified by Bullen (1975). If we use the present-day values of γ and Δ and if we add the centrifugal component into the second term of Eq. (3.1b), we even obtain remarkably similar values compared to GRS80 calculations (Moritz, 1988; see also Chapter 4, Normal Earth Gravity Field Versus Gravity Effect of

Layered Ellipsoidal Model, Table 3.). Surprisingly, one can find a reference to the fact that Bouguer's second term represents the free-air correction in the literature only scarcely (e.g., Putnam, 1895; Bullen, 1975, as quoted earlier), while the vast majority of recent textbooks do not mention it.

3.4.1.5 Bouguer's Third Term

Retrospectively, we should consider Bouguer's estimation of the gravitational effect of the topographic masses by the third term of Eq. (3.1) or (3.1b) as excellent. The quantity we now call the "Near Topographic Effect" (*NTE*, as a counterpart to the Distant Topographic Effect or *DTE*, Mikuška et al., 2006; see also Chapter 5, Numerical Calculation of Terrain Correction within the Bouguer Anomaly Evaluation (Program Toposk) can be well approximated by this term as illustrated in Fig. 3.4. In reality the real *NTE* is

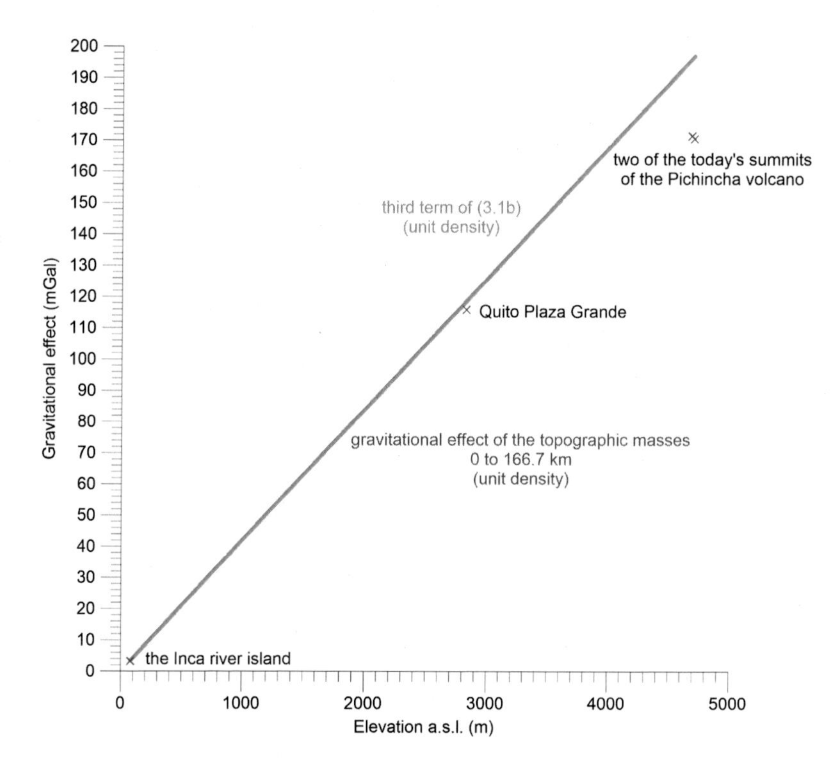

Figure 3.4 The estimation of the gravitational effect of the topographic masses by the third term of the expression (3.1b) (red line(dark gray in print versions)) and its values calculated from the available DEM models (blue crosses(light gray in print versions)) using the program Toposk (Marušiak et al., 2013; see also Chapter 5, Numerical Calculation of Terrain Correction within the Bouguer Anomaly Evaluation (Program Toposk).

proportional to h in the sense that Bouguer's third term represents approximately its upper limit. The third term of Eq. (3.1b) coincides well with the actual *NTE* calculated for Quito, i.e., the place where this approximation had been introduced.

3.4.1.6 Some of the Reasons Why Bouguer's Density Estimation of the Topographic Masses Was Unrealistic

Bouguer's density estimation was in fact unconvincing. We regard three reasons as having special importance: (1) There was poor general knowledge about the real rock densities or the density structure of the Earth in the half of the 18th century, and, instead, there were lots of "density speculations." (2) The gravitational constant had not yet been recognized or measured and, as a result, Pierre Bouguer could not confront the real full-valued differences between the measured and expected pendulum lengths with his prediction to which he had arrived at by combining the three terms of expression (3.1); obviously he only could calculate fractions. (3) Possibly even more fundamental, he was not aware that subtopographical lateral density changes can quite dramatically change the gravitational effect which he calculated on his approximation sphere surface. He simply supposed (in fact he had to suppose) that the spherical part of the real Earth would always behave like the spherical part of his "normal Earth" model, i.e., as if it were equigravitational. Unfortunately, this was rather far from geological reality. To avoid the effects of those lateral density changes, Airy (1856), von Sterneck (1883), and subsequently many others, tended to move into vertical mining shafts with their measurements aimed at the density estimations.

3.4.2 Faye (1880–95)

While we consider Faye's association with the introduction of the free-air correction to be slightly controversial, Faye deserves full credit for his other contributions to the gravity method.

Faye (1880b, p. 1446) uses the word "anomalies" in a direct relation to gravity values (more accurately to the pendulum lengths). We can then trace back the concept (and the term) of "gravity anomaly" as far as to 1880. In his 1883 memoir (Faye, 1883, p. 1261) he uses the term "Poisson's correction" for what Helmert (1884) recommended using "Bouguer reduction" as we already wrote earlier. Faye (1880b) seems

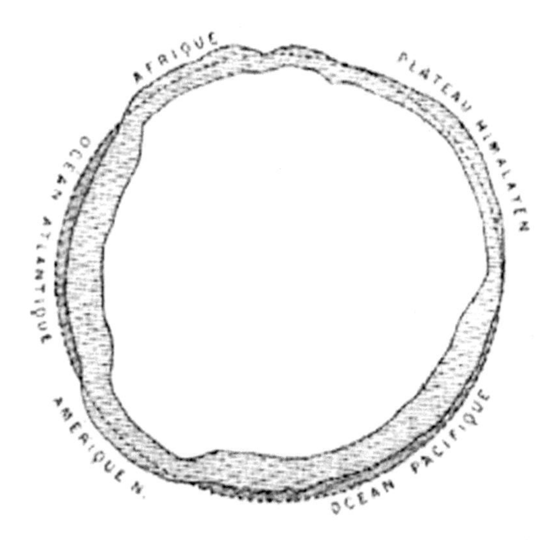

Figure 3.5 Faye (1895) obviously expected thicker crust under oceans and thinner under mountain ranges (his Figure 2 on his page 1085).

to have firstly dealt with modern isostasy concepts, even if he ends up with wrong conclusions later on (Faye, 1895; Fig. 3.5).

3.4.3 Helmert (1884)

Without any doubt Friedrich Rudolf Helmert was an outstanding scientist with enormous influence on the contemporary as well as the next generations of geodesists. This is well demonstrated by Putnam (1895, pp. 43, 44, 46, 56) or Hayford and Bowie (1912, p. 12), for example. However, if we are to consider Helmert's contribution especially from the perspective of the history of the Bouguer anomaly, we find that it should be viewed rather questionable.

There was, for instance, the way how he described condensations and reductions. Regarding the latter, Helmert (1884, p. 162) wrote: "The measurements of the acceleration of gravity actually apply for points at the physical Earth surface and thence they have to be reduced to the sea level in order they could be considered as belonging to one level plane." Such description then would logically have its impact not only upon geodesists but later on also upon geophysicists, for decades to come after publication of the Helmert's textbook. Although Helmert probably meant the quoted phrase as related to the "geoid computations," likewise it could easily be (mis)understood as "valid generally." However the same time it should be stressed here that not

Helmert (1884), but geophysicists themselves have in fact been responsible for the negative interpretational impact of the idea of "belonging to one level plane."

We perceive the notion of reductions very important from the aspect of our topic and we believe the following comment will not be redundant. Helmert (1884, p. 166) provides the expression

$$\left(1+\frac{2h}{r}\left[1-\frac{3}{4}\frac{\delta}{\Delta}\right]\right)g. \tag{3.8}$$

In Eq. (3.8) we exchanged his original symbols by those used by Bouguer (1749) (see expression (3.1)), except for the symbol g which is new in Eq. (3.8) and represents, according to Helmert, the measured gravity. It is clear that Eq. (3.8) was intended to describe "the traditional reduction of the gravity measurements to the sea level" (Helmert, 1884, p. 163). Here Helmert proposes to call the reduction after Bouguer as we mention before. It is, however, trivial to realize that if we multiply Eq. (3.8) by the factor $r\Delta/g$ we get Bouguer's expression (3.1) with changed signs of both the second and third terms. But then there could be a logical question: Provided that Eq. (3.1) is considered (formally) correct, can Eq. (3.8) be considered (formally) correct as well? We think it cannot be. The reason is that multiplication by variable g in Eq. (3.8) would mean that the change of gravity should depend on (variable) gravity itself, which would hardly have a physical reason especially regarding the third term. At least Helmert (1884) did not mention anything that could be considered a physical explanation.

Helmert (1884, p. 179) discusses a topic which has something in common with what we would call today the bathymetric correction. Helmert's reasoning is interesting but it is rather difficult to follow since it is amalgamated with his condensations. Further, on his pages 227–229 he criticizes the approach of Faye (1880b, p. 1444, see our discussion regarding Eq. 3.5) to the problem of compensation of local topographical features. This criticism of Helmert we consider substantiated since Faye (1880b, p. 1444) did not very much care about detailed specifications.

3.4.4 Putnam (1895)

Putnam (1895, p. 43) reproduces the formula, which was given by Helmert (1884) (our expression 3.8) when describing the process of "reducing pendulum observations to the sea level," and calls it Bouguer reduction according to Helmert (1884). On the other hand, Putnam considers *g* as gravity at sea level, contrary to Helmert. This is interesting. Here Putnam is closer to Bouguer (1749) than to Helmert (1884).

Further Putnam (1895, pp. 43, 44) applies the corrective term for the situation "whenever the topography about a station departs materially from this condition" and calls it "topographical correction" which should be always positive as he writes. Here Putnam quotes Helmert (1884, p. 169). We can only add that the concept of terrain correction (today the term "terrain" is preferred rather than "topographic") was probably introduced 40 years earlier by Peters (1855, p. 46).

Putnam (1895, pp. 53, 54) uses, discusses, and criticizes the Faye's method or Faye's reduction (1880b) while later Faye (1895) calls it correction. Please note that this method (reduction) of Faye is different from the free-air or Faye's reduction (correction) as we refer to elsewhere.

3.4.5 Hayford and Bowie (1912)

It is obvious that both the Earth model below sea level (including sea water and compensation of topographic masses) and the topographic masses themselves contribute to the theoretical gravity of Hayford and Bowie (1912).

In addition, it should be stressed that their approach was revolutionary also from other aspects as we briefly mentioned earlier. Hayford and Bowie (1912, p. 72) wrote: "It would be difficult to show satisfactorily by pure theory without numerical values why and to what extent the *curvature* and *distant topography* and compensation must be considered. ... Instead the computations have been made to cover the whole earth by formulae which are practically exact, curvature being adequately taken into account. This having been done the numerical results ... demonstrate conclusively and clearly that both distant topography and curvature must be considered if one is to secure even a fair approximation to the truth."

In the last quotation we did not highlight compensation since it should be considered an assumptive issue while the other two, namely the earth curvature and distant topography, were undoubtedly pure reality. In fact they combined together compensation and topographic effects, i.e., the assumptive and nonassumptive components, and that was unnecessary. We think it would have been much more logical if they had accounted for the topography and its compensation separately.

Among the constants introduced by Hayford and Bowie (1912) we can recognize three with the greatest importance, namely (1) the density of the near-surface crust: 2670 kg/m^3, (2) the outer limit of the near topographical masses with regard to a gravity station: 166.7 km, and (3) the depth of compensation: 113.7 km.

1. "... the mean density of the solid portion of the earth for the first few miles below the surface is assumed in this investigation to be 2.67" (i.e., 2670 kg/m^3; Hayford and Bowie, 1912, p. 10). More details about the background of this value, its adoption in geophysics as well as how Hayford (1909) and Hayford and Bowie (1912) worked with it is given in Hinze (2003). As a matter of fact, the assumption that the topographic masses have the density of 2.67 (something like a global average) has outlasted until today, although some different values have been used regionally, and although the concept of variable densities has been introduced in the meantime (Vajk, 1956, and many others).
2. The distance of 166.7 km from the gravity station (the outer limit of their zone O) was introduced by Hayford and Bowie (1912) as the boundary between their near and distant zones but they did not explain their choice. This figure (166.7 km) has also been in use until today. Bullard (1936) in fact introduced its use as an outer limit for allowing for topography, although Hayford and Bowie (1912) recommended to calculate the topographical effect around the Earth.
3. Regarding their compensation depth we can quote: "In the computations of the investigation here published the depth of compensation is assumed to be 113.7 kilometers under every separate portion of the earth's surface" (Hayford and Bowie, 1912, p. 10). Unlike the other two of their figures mentioned earlier, this compensation depth already does not seem to be "on duty." It is likely that this has happened at least partly because the Pratt–Hayford isostatic system as such has been generally less and less in use since there have been strong arguments against it (e.g., Glennie, 1932, p. 26 and elsewhere; Evans and Crompton, 1946, p. 215 and elsewhere).

Within the editorial discussion devoted to the last quoted paper one can find even the following statements. "... the speaker hoped that it would soon be possible to say without shock to the followers of Hayford and his apostle, Bowie, that large areas were not in isostatic equilibrium ..." (Holland in Evans and Crompton, 1946, p. 246). Or ... "the attention given to Pratt's hypothesis (perhaps because it lent itself more readily to mathematical treatment) was unfortunate, and that it would have been better if it had never been put forward" (Evans in Evans and Crompton, 1946, p. 249). And, in addition, the recent independent studies like CRUST1.0 (http://igppweb.ucsd.edu/~gabi/crust1.html) do not generally support the density distribution which would be required if such an isostatic system would have worked. In this light the statement of Hinze et al. (2013, p. 139), namely that "... in some cases the forces derived from these mass variations exceed the elastic limit of the lithosphere, leading to localized isostatic compensation as suggested by the Pratt-Hayford hypothesis" would require further specification.

3.4.6 Bullard (1936)

From the aspect of methodology (and history of our subject of course) the paper of Bullard (1936) represents something between what "had been" and what "has been." It is generally regarded as a milestone although in fact it brought little of what could be considered new from the aspect of methodology except for a few small changes or improvements.

Among others, Bullard criticized the way how Hayford and Bowie (1912) calculated the attraction of the topography between the station and their zone O. On his page 487 he wrote: "The work may be very much reduced if the attraction of a plateau on whose surface the station lies and which stretches to zone O is first calculated, and the difference between this and the attraction of the actual topography calculated by means of tables." Bullard did not mention, however, that a similar two-step method had already been proposed by Helmert (1884, pp. 169–172) and Putnam (1895, pp. 43, 44). The mentioned tables were those published by Cassinis and Doré (1933). Bullard calculated the attraction of the compensating masses within the inner zone according to the tables published by Heiskanen in 1931 and 1932. For saving work time Bullard sometimes combined the approaches of Heiskanen and Hayford. He considered the attraction of topography and compensation beyond zone O similarly as Hayford and Bowie (1912) by calculating the combined effect for all but one of his 87 pendulum stations.

To eliminate the evident discrepancy between the then used horizontal infinite slab and the real (nearly spherical) Earth he introduced his "slab extending to the outside of zone O and curved to the radius of the earth" (Bullard, 1936, p. 487) which we would now call truncated spherical layer or shell, with thickness equal to the station height. Bullard introduced an auxiliary term which, if added to the gravitational effect of a horizontal infinite slab, changes it into the effect of a "curved slab." This term now bears Bullard's name, being called either "Bullard term" or "Bullard B" (LaFehr, 1991). However, all this can be viewed as a rather complicated issue. Bullard (1936), along with many others later on, does not seem to have acknowledged that Bouguer (1749), when allowing for the (near) topographic masses, in fact did account for the Earth curvature, considering his "half of the effect which would be produced by the complete spherical layer" as we quote him earlier when discussing expression (3.3). So there is an obvious difference between the approaches of Bullard (1936) and Bouguer (1749), respectively, which we would like to point out here. We mean that while Bullard's maximum angular distance or outer limit (the one adopted from Hayford and Bowie, 1912, namely $1°29'58''$) remains always constant and is independent on the height of the actual measuring point h, Bouguer's "maximum angular distance" had to be dependent on h, and thus varying from point to point. Let us recall that Bouguer, for his approximation of the (near) topographic masses, did use neither infinite plate nor (truncated) spherical layer but spherical cap. However, it should be noted here that Bouguer (1749) had not discussed the question of the maximum angular distance or outer limit explicitly. Nevertheless, until today, the procedure has remained the same as it was introduced by Bullard (1936).

Interestingly from the historical aspect, Bullard (1936) quotes Helmert only seldom. This happens on his page 487 and here it is associated to the earlier gravity work in East Africa prior to Bullard's measurements, and then on his page 501 and following in connection with Helmert's formula for γ_0 which is identical with the one which we reproduce as our expression (3.7) except for the used equatorial value. On the contrary, Hayford and Bowie (1912) quote Helmert more frequently.

And Putnam (1895) did so yet more often. On his page 56 he even quotes a passage from Helmert (1890) in German which we consider very illustrative. With some caution we can interpret this matter as a slight shift from more or less geodesy to more or less geophysics, within approximately one or two human generations of the elapsed time.

3.4.7 Abandoning Reductions to the Sea Level

After the works of Hayford and Bowie (1912) and also Bullard (1936) were published, one would expect that those questionable reductions would either disappear from the geophysical literature or, at least, their occurrence would be less and less frequent. Unfortunately, this has not been the case.

For instance Heiskanen and Vening Meinesz (1958, p. 147) wrote: "Before we use the observed gravity values for practical purposes or for theoretical studies, we must reduce them in a proper way to the same level, usually to sea level or the geoid." We interpret this statement as a misunderstanding of the authors who had been, no doubt, two of the topmost specialists.

On the other hand the approach that gravity anomaly should be understood as a station anomaly was outlined and stressed for instance by Grant and Elsaharty (1962, p. 616), Tsuboi (1965, pp. 386, 387), and Naudy and Neumann (1965, pp. 2, 3).

LaFehr (1991) drew attention to this problem again and did what was possible to explain and to demonstrate the fatal error we can commit when not avoiding this kind of reductions. He wrote (LaFehr, 1991, p. 1177): "How widespread this notion is can be seen by reviewing the major textbooks on the subject: of 15 English-language books which carry descriptions, no fewer than nine state or imply" ... "that our intent is datum reduction." He then quotes three of the textbooks giving correct explanations of what he called "the data reduction process."

Later on Li and Götze (2001, p. 1660) wrote, feeling that there was still some confusion among geophysicists regarding the sense of the free-air reduction: "However, the 'free-air' reduction was thought historically to relocate gravity from its observation position to the geoid (mean sea level). Such an understanding is a geodetic fiction, invalid and unacceptable in geophysics."

More recently the fact that gravity anomaly should be related to the gravity station was stressed among others by Hinze et al. (2005, p. J28), to quote the publication, which has been coauthored by as many as 21 prominent, mostly North-American, geophysicists.

However, even today the confusion is still present. For example, although referring to Hinze et al. (2005), Mallick et al. (2012, p. 5) wrote: "When the gravity observations are made at two stations, each located at different elevations, there would be a difference in the two gravity readings at these stations, which if not corrected for, might indicate a spurious sub-surface structure. This variation in the gravity measurements can be removed by introducing a datum plane with a certain elevation above sea level. *All the material above the datum plane is mathematically removed so that the instrument can be effectively imagined to be placed on top of the datum surface.*"

Unfortunately Mallick et al. (2012) are not the only ones who recently published an incorrect approach to gravity reductions in applied (exploration) geophysics. Long and Kaufmann (2013, pp. 25, 29) represent another example, although they also refer to Hinze et al. (2005). On the other hand, for instance, the approach of Dentith and Mudge (2014, pp. 103, 104) is correct.

In the light of the above mentioned the only possible outlook from our historical hindsight is that the struggle against reductions to the sea level should continue.

3.4.8 Just Three Terminological Comments

Above all we would like to discuss the term "reduction" or "reduce" which has been quite frequently used in a variety of meanings. For instance in Bullard (1936, p. 450) that term was used for lessening the number of some specific measured quantities. Further Bullard (1936, p. 487) writes about the tables of Hayford and Bowie (1912): "... first the attraction of the topography in the compartment on a station at the same height as the mean level of the compartment is taken from the main table, then the correction necessary to reduce this to the attraction at the actual height of the station is found from a subsidiary double entry table" and immediately after he uses that term for description of saving work. A different meaning can be found in Putnam (1895, p. 52): "... Apply further correction to the observed force of gravity ... to reduce to the normal condition." Bouguer (1749) uses this term in connection with projection of the measured distances on the sea level as we quote earlier . . . Young, Stokes, Faye, Helmert, and many others used this term when describing the relocation of gravity values.

From our brief (and thus necessarily incomplete) review it follows that the term "reduction" has been in use not only in the sense of (from the geophysical point of view invalid) moving gravity values along the local vertical but also in a number of other senses. We thus consider it a most problematic and troublesome terms in gravimetry.

Then there is "Bouguer anomaly." In Bullard (1936) that term has been used more or less in the sense as we use it today. In Hayford and Bowie (1912), however, Bouguer anomaly is understood as a quantity defined on a flat Earth. Putnam (1895) calls the same quantity "residuals with Bouguer reduction" (together with "residuals with reduction for elevation" in the case of free-air anomaly).

And finally there is the term "free-air." For instance either Stokes (1849) or Faye (1880a,b, 1883) had not used this term for the reduction in question while Helmert (1884, p. 166) had (his term "in freier Luft"). It can be of some interest that the term "free-water anomaly" was introduced one century later (Luyendyk, 1984), as an apparent analogy to the "free-air anomaly." It may be important to note, however, that the term "free-air" should be understood in the sense "as if there were air while in fact there was rock," not "there was in fact air and no rock." In this light Luyendyk's "analogy" does not seem very fortunate. Or, is not the term "free-air" also trouble-causing?

3.5 CONCLUSIONS

From the present-day aspect, the development of the Bouguer anomaly concept seems to be a never-ending story. As if it was an object of a complicated evolution rather than a result of some purposeful plan or idea. In addition, it has been closely entwined with geodetic concepts which have proved to be geophysically unintelligible.

As we already indicated earlier, we have decided to terminate our historical excursion approximately at the time of appearance of the paper of Bullard (1936). And we have learned that, until then, some concepts developed differently than we thought we know today.

For instance Bouguer realized already in 1749 that both "mass or rock" and "free-air" components affect the gravity change when the observer moves from lower to higher elevations. Today we usually call those quantities as corrections. The former bears Bouguer's name probably thanks to Helmert who used it for relocating the measured gravity values what Bouguer in fact neither did nor proposed. The latter is either called "free-air" or "Faye" correction. The idea undoubtedly came from Bouguer although Bouguer (1749) never used the term "free-air" which, in turn, has been probably coined by Helmert (1884). And the idea of "free-air" upward diminution of gravity definitely did not originate from Faye.

Faye, on the other hand, promoted the isostatic compensation of the topographic masses in 1880 which was earlier than Fisher (1881) and Dutton (1882, 1889) published their works. For example, Fisher and Dutton are quoted in Watts (2001, pp. 15–17), while Faye is not. Faye, however, did not use the term "isostasy." In this aspect Faye (1880b) seems to have been omitted not only by Watts (2001); in fact we have not found his isostatic thoughts mentioned anywhere.

Bouguer in fact did not introduce the "Bouguer slab" or "Bouguer plate" as we showed earlier.

Although it has been well known that Hayford and Bowie (1912) changed the sense of applying the Bouguer and free-air (sometimes called Faye) corrections on the theoretical instead of the measured gravity, unfortunately this change has not been fully acknowledged. As we have showed, the related misunderstandings have not yet been eliminated.

At the very end of our retrospective we have to highly appreciate once more the contribution of Pierre Bouguer who, as early as it was possible, learned much of what was necessary in order to establish the concept of the gravity anomaly. He introduced a simple normal Earth consisting of an equigravitational sphere plus topography and, on this basis he was capable of predicting gravity at elevated places compared to its value near the sea level. He well realized the role of centrifugal force and distinguished between gravitation and gravity. On the other hand, he seems to have disregarded the possible influence of bathymetry and, what was in our view even more important, he in his time had no means how to interpret the difference between the measured and expected pendulum lengths in terms of lateral density changes within the subtopographical volume of the real Earth. This led him to questionable topographic density estimation. However, we can conclude that Pierre Bouguer had laid the foundations of the present-day Bouguer anomaly house. Yet in his times he could not raise the walls, since not all of the necessary bricks he had in his hands.

ACKNOWLEDGMENTS

This work was supported by the Slovak Research and Development Agency under the contracts APVV-0194-10 and APVV-0827-12. We thank John Stanley, Bruno Meurers, and Hans-Jürgen Götze for their help with editing the manuscript.

REFERENCES

Airy, G.B., 1856. Account of pendulum experiments undertaken in the Harton Colliery for the purpose of determining the mean density of the Earth. Philos. Trans. R. Soc. Lond. 146, 297–355.

Bouguer, P., 1749. La figure de la terre. Charles-Antoine Jombert, Paris, 394 pp.

Bullard, E.C., 1936. Gravity measurements in East Africa. Phil. Trans. R. Soc. Lond. A 235, 445–534.

Bullen, K.E., 1975. The Earth's density. Chapman and Hall, London, 420 pp.

Cassinis, G., Doré, P., 1933. Tables fondamentales pour les réductions des valeurs observées de la pesanteur: Éditions provisoire presentée à la Commission internationale de la pesanteur. Lisbonne, 1933.

Chapman, M.E., Bodine, J.H., 1979. Considerations of the indirect effect in marine gravity modeling. J. Geophys. Res. 84 (B8), 3889–3892.

Cox, A.N. (Ed.), 2002. Allen's Astrophysical Quantities. Springer + Business Media, New York, 719 pp.

Dentith, M., Mudge, S.T., 2014. Geophysics for the Mineral Exploration Geoscientists. Cambridge University Press, Cambridge, 438 pp. plus Appendices.

Dutton, C.E., 1882. Physics of the Earth's crust; by the Rev. Osmond Fisher. Am. J. Sci. 23 (136), 283–290.

Dutton, C.E., 1889. On some of the greater problems of physical geology. Bull. Philos. Soc. Washington 11, 51–64.

Ecker, E., Mittermayer, E., 1969. Gravity corrections for the influence of the atmosphere. Bolletino di Geofisica Teorica ed Applicata 11 (41 and 42), 70–80.

Evans, P., Crompton, W., 1946. Geological factors in gravity interpretation illustrated by evidence from India and Burma. Quart. J. Geol. Soc. 102, 211–249.

Faye, H.A.É.A., 1880a. Sur les variations séculaires de la figure mathématique de la Terre, 90. Comptes Rendus des Séances de l'Académie des Sciences, Paris, pp. 1185–1191.

Faye, H.A.É.A., 1880b. Sur la réduction des observations du pendule au niveau de la mer, 90. Comptes Rendus des Séances de l'Académie des Sciences, Paris, pp. 1443–1446.

Faye, H.A.É.A., 1883. Sur la réduction du baromètre et du pendule au niveau de la mer, 96. Comptes Rendus des Séances de l'Académie des Sciences, Paris, pp. 1259–1262.

Faye, H.A.É.A., 1895. Réduction au niveau de la mer de la pesanteur observée à la surface de la Terre par M.G.R. Putnam, 120. Comptes Rendus des Séances de l'Académie des Sciences, Paris, pp. 1081–1086.

Fisher, O., 1881. Physics of the Earth's crust. Macmillan and Co., London, 299 pp.

Glennie, E.A., 1932. Gravity anomalies and the structure of the Earth's crust: Survey of India Professional Paper No. 27. Dehra Dun, 35 pp.

Grant, F.S., Elsaharty, A.F., 1962. Bouguer gravity corrections using a variable density. Geophysics 27, 616–626.

Hayford, J.F., 1909. The Figure of the Earth and Isostasy from Measurements in the United States. Department of Commerce and Labor, Coast and Geodetic Survey, Special Publication No. 82, 178 pp.

Hayford, J.F., Bowie, W., 1912. The Effect of Topography and Isostatic Compensation Upon the Intensity of Gravity. U.S. Coast and Geodetic Survey, Special Publication No. 10, 132 pp.

Heiskanen, W.A., Vening Meinesz, F.A., 1958. The Earth and Its Gravity Field. McGraw – Hill Book Company, New York, 470 pp.

Helmert, F.R., 1884. Die matematischen und physikalischen Theorieen der Höheren Geodäsie. Teil II, Teubner, Leipzig, 610 pp.

Helmert, F.R., 1890. Die Schwerkraft im Hochgebirge, insbesondere in den Tyroler Alpen in Geodätischer und Geologischer Beziehung. Veröffentlichung des Königlichen Preussischen Geodätischen Institutes, pp. 1–52.

Hinze, W.J., 2003. Short note: Bouguer reduction density, why 2.67? Geophysics 68, 1559–1560.

Hinze, W.J., Aiken, C., Brozena, J., Coakley, B., Dater, D., Flanagan, G., et al., 2005. New standards for reducing gravity data: The North American gravity database. Geophysics 70, J25–J32.

Hinze, W.J., von Frese, R.R.B., Saad, A.H., 2013. Gravity and Magnetic Exploration. Cambridge University Press, Cambridge, p. 512.

LaFehr, T.R., 1991. Standardization in gravity reduction. Geophysics 56, 1170–1178.

Lambert, W.D., 1930. The reduction of observed values of gravity to sea level. Bulletin Géodesique 26, 107–181.

Li, X., Götze, H.-J., 2001. Ellipsoid, geoid, gravity, geodesy and geophysics. Geophysics 66, 1660–1668.

Long, L.T., Kaufmann, R.D., 2013. Acquisition and Analysis of Terrestrial Gravity Data. Cambridge University Press, Cambridge, p. 171.

Luyendyk, B.P., 1984. On-bottom gravity profile across the East Pacific Rise crest at 21° north. Geophysics 49, 2166–2177.

Mallick, K., Vashanti, A., Sharma, K.K., 2012. Bouguer Gravity Regional and Residual Separation: Application to Geology and Environment. Springer & Capital Publishing Company, New Delhi, 288 pp.

Marušiak, I., Zahorec, P., Papčo, J., Pašteka, R., Mikuška, J., 2013. Toposk, program for the terrain correction calculation: G-trend, s.r.o., Bratislava, unpublished manual (in Slovak).

Mikuška, J., Pašteka, R., Marušiak, I., 2006. Estimation of distant relief effect in gravimetry. Geophysics 71 (6), J59–J69.

Moritz, H., 1988. Geodetic reference system 1980. Bull. Géod. 62 (3), 348–358.

Naudy, H., Neumann, R., 1965. Sur la definition de l'anomalie de Bouguer et ses consequences pratiques. Geophys. Prospect. 13, 1–11.

Peters, C.A.F., 1855. Die Länge des einfachen Secundenpendels auf dem Schlosse Güldenstein. Astron. Nachr. 40 (937–945), 1–152.

Petit, G., Luzum, B. (Eds.), 2010. IERS Conventions (2010). Verlag des Bundesamts für Kartographie und Geodäsie, Frankfurt am Main.

Poisson, S.D., 1833. Traité de mécanique (Seconde édition, Tome premier). Bachelier, Imprimeur – Libraire, Paris, pp. 492–496.

Putnam, G.R., 1895. Results of a transcontinental series of gravity measurements. Bull. Philos. Soc. Washington 13, 31–60.

von Sterneck, R., 1883. Wiederholung der Untersuchungen über die Schwere im innern der Erde, ausgefürht im Jahre 1883 im dem 1000 m tiefen Adalbertschachte des Silberbergwerkes zu Přibram in Böhmen, 3. Mittheilungen des Kaiserliche-Königliche Militär-Geographischen Institutes zu Wien, pp. 59–94.

Stokes, G.G., 1849. On the variation of gravity at the surface of the Earth. Trans. Cambridge Philos. Soc. 8, 672–695.

Todhunter, I., 1873a. A History of the Mathematical Theories of Attraction and the Figure of the Earth from the Time of Newton to that of Laplace, Volume I. MacMillan and Co., London, 476 pp.

Todhunter, I., 1873b. A History of the Mathematical Theories of Attraction and the Figure of the Earth from the Time of Newton to that of Laplace, Volume II. MacMillan and Co., London, 508 pp.

Tsuboi, C., 1965. Calculations of Bouguer anomalies with due regard the anomaly in the vertical gradient. Proc. Jap. Acad. B 41 (5), 386–391.

Vajk, R., 1956. Bouguer corrections with varying surface density. Geophysics 21, 1004–1020.

Watts, A.B., 2001. Isostasy and Flexure of the Lithosphere. Cambridge University Press, Cambridge.

Young, T., 1819. Remarks on the probabilities of error in physical observations, and on the density of the earth, considered, especially with regard to the reduction of experiments on the pendulum, In a letter to Capt. Henry Kater, F.R.S. Philos. Trans. Roy. Soc. Lond. 109, 70–95.

CHAPTER 4

Normal Earth Gravity Field Versus Gravity Effect of Layered Ellipsoidal Model

Roland Karcol[1,2], Ján Mikuška[3] and Ivan Marušiak[3]
[1]Comenius University, Bratislava, Slovak Republic [2]Slovak Academy of Sciences, Bratislava, Slovak Republic [3]G-trend, s.r.o., Bratislava, Slovak Republic

4.1 INTRODUCTION

Loosely speaking, Bouguer anomaly is the measured gravity corrected for the known or modeled gravity effects at a planetary scale. The first of these corrections is subtracting the normal field or normal gravity. One can find several, slightly different, definitions of the normal field. For instance, the one by Jacoby and Smilde (2009) states, "Normal field would be observed on the oblate reference ellipsoid of rotation that approximates the equilibrium figure best fitted to Earth."

The well-known formula of Somigliana (1929) is commonly used to calculate the normal field:

$$\gamma = \gamma_E \frac{1 + k\sin^2\varphi}{\sqrt{1 - e^2\sin^2\varphi}} \tag{4.1}$$

where $k = \left(b\gamma_P/a\gamma_E\right) - 1$, $e^2 = \left(a^2 - b^2\right)/a^2$, a is the major semiaxis, b is the minor semiaxis, γ_P and γ_E are theoretical gravity values at the equator and at the poles, respectively, e is the first eccentricity, and φ is the latitude. This formula is based on very important condition of constant potential on the surface of the ellipsoid. No information about density distribution within the model is known nor required (there are infinitely many density distributions to produce the same total mass). The (normal, reference) ellipsoid is defined by four parameters: major semiaxis a, flattening f, angular velocity ω, and GM value (product of the gravitational constant G and the mass of the ellipsoid M). These parameters are based on satellite measurements. The values of γ_P and γ_E are calculated from defining parameters.

Understanding the Bouguer Anomaly. DOI: http://dx.doi.org/10.1016/B978-0-12-812913-5.00003-8

4.2 SOME PROBLEMS WITH THE NORMAL FIELD

We want to discuss two problems linked with using Eq. (4.1). The first one is the mass of the model. The surface of the ellipsoidal model represents the zero level of altitude, but its mass, represented by the *GM* value incorporated in Eq. (4.1), contains both the masses of topography and atmosphere. That means, these masses have been "pushed" into the ellipsoidal model interior. However, any moving of the masses from their correct or real positions is in general inadmissible in applied geophysics/gravimetry. The problem of topography is much more complicated, and it would require intensive and careful studies (and lots of hard discussions, because the "audience" is demanding), so it will be omitted in our model for now. On the other hand, the *GM* value with Earth's atmosphere excluded is presented in the specification NIMA (2000) and is used in presented calculations.

The second problem is the density distribution within the model. Formula (4.1) is based on idea of a constant potential on the surface. Yet the density distribution must be specific in order to produce a constant potential along the surface (see Fig. 4.1 and e.g., Moritz, 1968). One can find overly simple two-layered example consists of homogenous core enclosed by nearly homogenous mantle in Moritz (1990). Two layers are, of course, not enough. However, in the literature, one

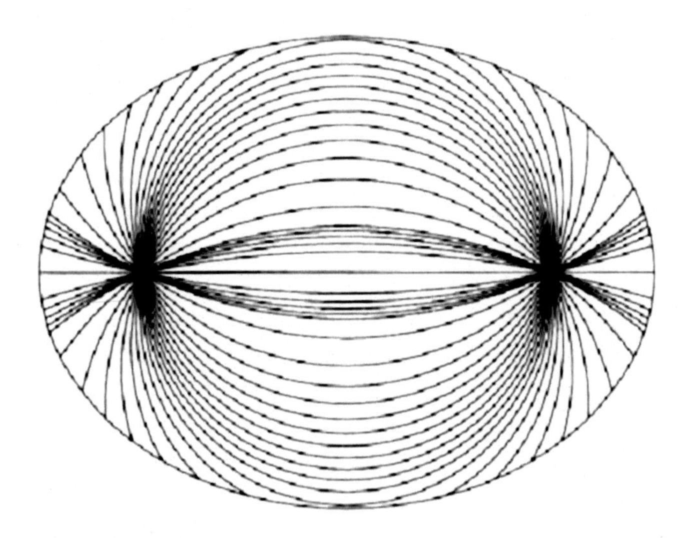

Figure 4.1 An example of one possible density distribution which would produce constant potential on the surface of ellipsoid, according to Conway (2000). Black lines stand for constant density surfaces which, in this cross-section, are represented by isodensity lines.

can find the statements which say that the question of density distribution is not important (e.g., Heiskanen and Moritz, 1967, p. 64). While this could be admissible for geodesy in the problem of the shape of the Earth, it is not acceptable for geophysics/gravimetry. Since the true inner structure of the Earth is close to a layered model (as we are convinced by seismic observations), the usefulness of the discussed condition of the constant potential should be critically revaluated.

4.3 THE GRAVITY POTENTIAL AND GRAVITY EFFECT OF THE EQUIPOTENTIAL MODEL

The first question is the magnitude of the difference between potentials and gravity effects of models with and without including the mass of the atmosphere. The defining parameters of the ellipsoidal model according to the specifications of NIMA (2000) are as follows:

- semimajor axis $a = 6378{,}137$ (m)
- flattening $f = 1/298.257223563$
- GM with the atmosphere included $GM = 3986{,}004.418 \cdot 10^8$ ($\mathrm{m}^3/\mathrm{s}^2$)
- angular velocity $\omega = 729{,}2115 \cdot 10^{-11}$ (rad/s).

The required derived parameters are as follows:

- semiminor axis $c = a(1 - f) \approx 6356{,}752.3142$ (m)
- GM with the atmosphere excluded $GM' = 3986{,}000.9 \cdot 10^8 \ (\mathrm{m}^3/\mathrm{s}^2)$

The value of gravitational constant G in this specification is: $G = 6.673 \cdot 10^{-11}\ \mathrm{m}^3/\mathrm{kg}$ per s^2 what is a little bit less than the most recent update according to Mohr et al. (2012) which reads: $G = 6.67384 \cdot 10^{-11}\ \mathrm{m}^3/\mathrm{kg}$ per s^2. However, we continue in using the NIMA value of G in order to stay consistent with defining constants (a, f, ω, GM) and because the last mentioned GM' value is still not common in other specifications of constants e.g., Petit and Luzum (eds.) (2010).

The formula for gravity potential, as a sum of the gravitational potential and the centrifugal component, in calculation points on its surface is (e.g., Heiskanen and Moritz, 1967, after some simplification)

$$U = \frac{GM}{e}\arctan\left(\frac{e}{c}\right) + \frac{1}{3}\omega^2 a^2, \tag{4.2}$$

where e is the linear eccentricity: $e = \sqrt{a^2 - c^2}$.

The value of potential for the GM value is $U = 62,636,851.7146\ \mathrm{m^2/s^2}$. The solution for the case without atmosphere is (GM' is used): $U = 62,636,796.4957\ \mathrm{m^2/s^2}$.

The gravity effects were obtained by using Eq. (4.1). First, we calculated the gravity values on the pole γ_P and on the equator γ_E. These values are derived from the defining parameters via the respective formulae (e.g., Heiskanen and Moritz, 1967):

$$\gamma_P = \frac{GM}{a^2}\left(1 + \frac{m}{3}\cdot\frac{e'q'}{q}\right), \tag{4.3}$$

$$\gamma_E = \frac{GM}{ac}\left(1 - m - \frac{m}{6}\cdot\frac{e'q'}{q}\right), \tag{4.4}$$

where $e' = e/c$ is the second eccentricity, $m = \left(\omega^2 a^2 c/GM\right)$, $q' = 3\left(a^2/e^2\right)\left[1 - \left(c/e\right)\arctan\left(e/c\right)\right] - 1$ and $q = 1/2\left[\left(1 + \left(3c^2/e^2\right)\right)\right.$ $\left.\arctan\left(e/c\right) - 3\left(c/e\right)\right]$. The formula of Somigliana (1929) (Eq. (4.1)) is then used to calculate the normal (theoretical) gravity as a function of φ. The results for GM (atmosphere included) and GM' (atmosphere excluded) are close to each other, and the difference between them is in the range from 0.863 to 0.868 mGal (Fig. 4.2).

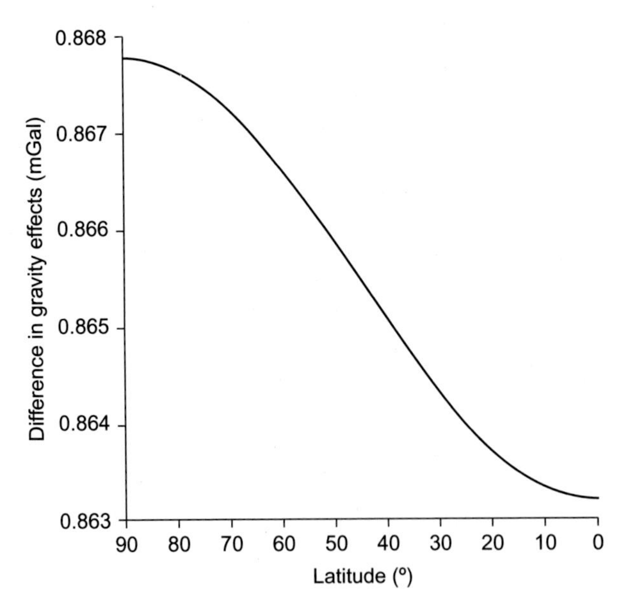

Figure 4.2 The difference in the normal (theoretical) gravity between the two discussed cases of the equipotential model (the Earth mass with and without atmosphere).

Please note that the commonly used atmospheric correction for $h = 0$ is 0.874 mGal which is a slightly higher figure than are those of our difference interval in Fig. 4.2, which is interesting. We can only speculate that this small-scale disagreement can be a result of the spherical approximation used by Wenzel while our values are based on the ellipsoidal model.

4.4 THE GRAVITY POTENTIAL AND GRAVITY EFFECT OF THE LAYERED MODEL

We attempted to address the previously mentioned problems (namely, the unwanted displacements of the existing masses, the realistic density distribution within the model, and the arbitrary position of the calculation point) by introducing the model with several layers in order to better approximation of the real density distribution within the Earth, without applying the limiting condition of a constant potential on the surface. There are, of course, some difficulties with this approach either. The first one is the shape of the layers. Generally, there are three possibilities how to construct the layered ellipsoidal model:

- to use a family of confocal ellipsoids,
- to take an advantage of a set of homeoids (linearly reduced ellipsoids), or
- to use ellipsoids with no specific mutual relations.

We have chosen the second option because of the important qualities of homeoids. The first one is the shape of the layers. The thickness of such layers is increasing from polar to equatorial areas and the eccentricity and sphericity of all layers remain the same (these properties do not hold for the confocal family). The second important advantage is the zero-gravitational effect within the inner hollow volume of a homeoid. This property is well known for the spherical shell, which represents, from this aspect, just a special case of homeoid. This property simplifies the calculations significantly.

The next problem is the density distribution itself. All the density tables we have come across were calculated for spherical approximation of the Earth (with the radius which provides equal volume to that of the ellipsoid). While the layers in our model are homeoids, we assigned the thicknesses of layers to the semimajor axis of the ellipsoidal model, and the thicknesses in semiminor axis were subsequently calculated. Unfortunately, the tabulated values of density are usually assigned to the upper and lower boundaries of each layer, and no exact information about density distribution within the layer is given, except of some graphical representations. As the first trial, we used the density table of Ochaba (1986) which is based on Bullen (1975), see Table 4.1. The density for each layer was set as the mean of the tabulated values for both the upper and the lower edge of individual layer, respectively. The density function within each layer could be, in the future work, partially replaced by a system of thin layers with a small density step (e.g., 0.01 g/cm^3 what is the common accuracy of the reported density values). It should be noted here that there are solutions for the gravitational potential of heterogeneous ellipsoid available, e.g., Dyson (1891) or Rahman (2001). These solutions are extraordinarily challenging, but they exceed the scope of this paper. Therefore, this possibility remains as another task for the future work for the time being.

The first step was to calculate the *GM* constant for this layered model and compare it with the tabulated one (atmosphere excluded). This task was reduced to calculate the mass of our model, while the constant *G* is known. We considered our first result $GM_{\text{layered}} = 3989,853.8 \cdot 10^8\ \text{m}^3/\text{s}^2$ too high, so that each density value in Table 4.1 has been reduced by a constant value of 0.00261 g/cm^3. The new result was $GM_{\text{layered}} = 3986,000.97 \cdot 10^8\ \text{m}^3/\text{s}^2$ what we regarded as a good fit to GM'. Finally, the input parameters for our

Table 4.1 Earth Density Distribution According to Ochaba (1986)

Depth (km)		Density (g/cm^3)	
Upper edge	**Lower edge**	**Upper edge**	**Lower edge**
0	33	2.84	3.32
33	245	3.32	3.51
245	984	3.51	4.49
984	2000	4.49	5.06
2000	2700	5.06	5.40
2700	2886	5.40	5.69
2886	4000	9.95	11.39
4000	4560	11.39	11.87
4560	4710	11.87	12.30
4710	5160	12.30	12.74
5160	6378.137	12.74	13.03

Table 4.2 Input Parameters for the Layered Ellipsoidal Model

Layers			
No.	Thickness (km)		Density (kg/m^3)
	Equatorial	Polar	
1	33	32.889	3002.39
2	212	211.289	3412.39
3	739	736.522	3997.39
4	1016	1012.594	4772.39
5	700	697.653	5227.39
6	186	185.376	5542.39
7	1114	1110.265	10,667.39
8	560	558.122	11,627.39
9	150	149.497	12,082.39
10	450	448.491	12,517.39
11	1218.137	1214.053	12,882.39

model are in Table 4.2. Please note that the density units were changed to kg/m^3.

Next, the gravity potential and gravity effect of the layered model are calculated. This is done in several steps (based on Kartvelishvili, 1982):

- Conversion of the calculation point coordinates to the Cartesian system. The conversion formulae are as follows:

$$x^2 = \left[\frac{a^2}{\sqrt{a^2\cos^2\varphi + c^2\sin^2\varphi}} + h\right]^2 \cos^2\varphi,$$

$$z^2 = \left[\frac{c^2}{\sqrt{a^2\cos^2\varphi + c^2\sin^2\varphi}} + h\right]^2 \sin^2\varphi,$$

  where h stands for altitude (positive, zero, or negative), φ is the latitude, and a and c are semimajor and semiminor axes, respectively. The calculation points are situated on the surface of the model ($h = 0$), and the step is 1° in latitude in the case of our example shown in Fig. 4.3.
- Calculation of the parameter λ, which “expresses” the change in the length of semiaxis. The gravitational effect of a given ellipsoid is

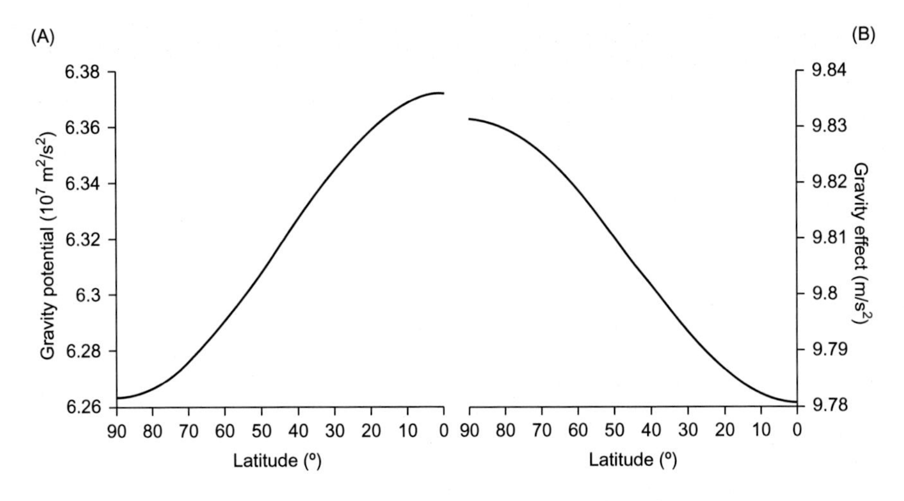

Figure 4.3 The gravity potential (A) and effect (B) of our layered ellipsoidal model.

calculated as the effect of confocal ellipsoid which passes through calculation point (if located outside of the given ellipsoid). If the original semiaxes are a and c, the semiaxes of the larger ellipsoid are $a' = \sqrt{a^2 + \lambda}$ and $c' = \sqrt{c^2 + \lambda}$. The formula for the parameter λ is as follows:

$$\lambda = \frac{x^2 + z^2 - a^2 - c^2 + \sqrt{e^4 - 2e^2(x^2 - z^2) + (x^2 + z^2)^2}}{2},$$

where $e = \sqrt{a^2 - c^2}$.

- Calculation of the parameter p: $p = \text{sign}(\lambda)\left(\frac{\text{sign}(\lambda)+1}{2}\right)\cdot\lambda = \begin{cases} \lambda & \text{if } \lambda > 0 \\ 0 & \text{if } \lambda \le 0 \end{cases}$.
- The centrifugal components: $W = 1/2\omega^2 x^2$ for the potential and $a_x = \omega^2 x$ for the effect
- The gravity potential of a single homogenous ellipsoid:

$$U = \frac{\pi G\sigma a^2 c}{e^3}\left[\left(2z^2 + 2e^2 - x^2\right)\text{arctg}\frac{e}{\sqrt{c^2+p}} + e\left(x^2\frac{\sqrt{c^2+p}}{a^2+p} - 2z^2\frac{1}{\sqrt{c^2+p}}\right)\right] + \frac{1}{2}\omega^2 x^2 \qquad (4.5)$$

where σ is the density, and G is the gravitational constant.

- Calculating of the x- and z-components of the gravitational attraction vector:

$$g_x = -\frac{2x\pi a^2 c\sigma G}{e^3}\left[\operatorname{arctg}\frac{e}{\sqrt{c^2+p}} - \frac{e\cdot\sqrt{c^2+p}}{(a^2+p)}\right] \tag{4.6}$$

$$g_z = \frac{4z\pi a^2 c\sigma G}{e^3}\left[\operatorname{arctg}\frac{e}{\sqrt{c^2+p}} - \frac{e}{\sqrt{c^2+p}}\right] \tag{4.7}$$

(e.g., MacMillan, 1930—integral form, or Kartvelishvili, 1982—closed form)

- The gravity effect of a single homogenous ellipsoid:

$$|\vec{g}| = \sqrt{(g_x + a_x)^2 + (g_z)^2} \tag{4.8}$$

The gravitational potential/effect of a single layer (homeoid) is obtained as a difference of the potentials/effects of two ellipsoids. The parameter p was added to the formula by us to avoid "if else" statements in our MATLAB code. These formulae allow us to calculate gravitational or gravity potential/effect of a homogenous ellipsoid for an arbitrary position of the calculation point (even inside the model). This is an obvious practical advantage in comparison with the formula of Somigliana (1929) which allows calculating these values only on the surface of the model. However, the formula for gravity potential and gravity effect of equipotential model extended to positive heights can be found in e.g., Li and Götze (2001), but still, such formulae do not work for negative heights.

The results for gravity potential and gravity effect are depicted in Fig. 4.3A and B.

The important thing is comparison of this solution with the effect of the homogenous ellipsoid with the same size and GM. Those effects are identical in the case of spherical model, and since the flattening of our layered ellipsoidal model can be regarded as insignificant ($f = 1/298.257223563 = 0.003352810664747$), one could expect similar solutions for ellipsoidal model, too. The result of this experiment is presented in Fig. 4.4. It is obvious that the difference is significant. The expected equality of effects will be true for models based on a confocal family of ellipsoids, but, as we already indicated above, such model is not suitable for modeling the layered structure of the real Earth.

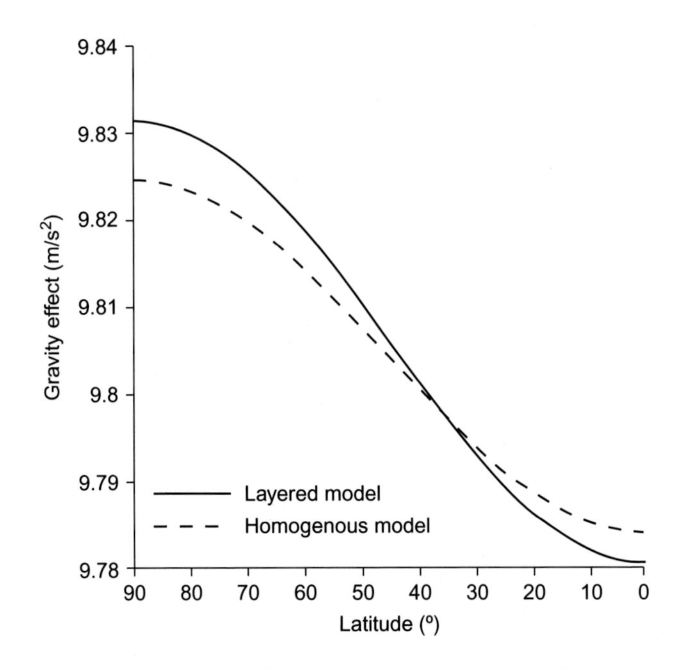

Figure 4.4 The comparison of gravity effects of the layered and homogenous ellipsoidal models.

4.5 COMPARISON OF THE RESULTS FOR THE EQUIPOTENTIAL AND THE LAYERED MODELS

The obtained results are depicted in Fig. 4.5A, while their difference is in Fig. 4.5B. This difference varies from −40.2 to 76.6 mGal. It is evident that there is nothing like a constant shift involved in this difference, what was, in fact, expected, and that the difference is not symmetric either around its zero or around 45° of latitude. The perfect fit is not possible, except of a situation when a thin layer with negative density is placed on the surface of layered model, as described in Pizzeti (1894), what is of course completely unrealistic. The goal was not to obtain the fit, but to show that a layered model can produce an acceptable output, with some previously mentioned advantages in comparison with the equipotential model.

If we take a closer look at the latitudes that limit e.g., Slovakia (from ~47.7°N to ~49.7°N), the discussed difference gets the shape of a linear trend with the range close to 4 mGal (Fig. 4.6). This will be true for many countries, while only the value of the range will be changed depending on the meridional extent of the particular country.

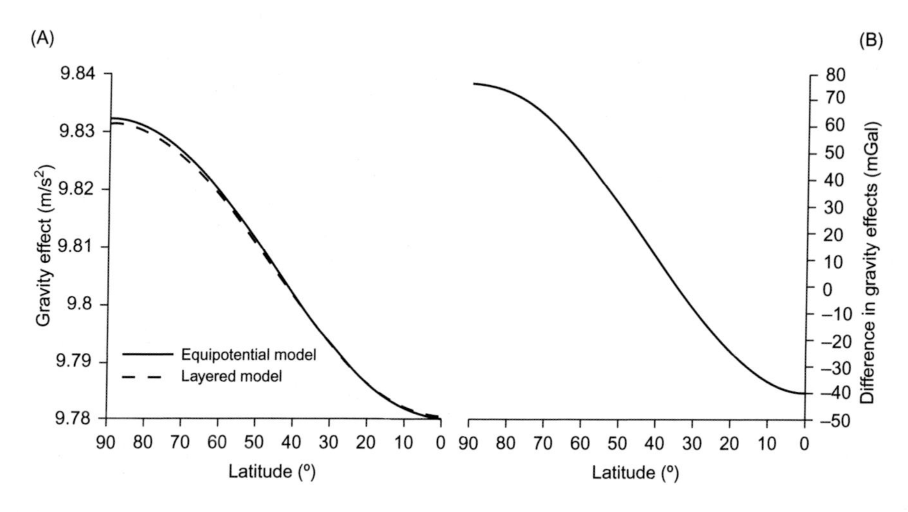

Figure 4.5 (A) The comparison of the solutions for equipotential and layered models; (B) the difference equipotential minus layered solution.

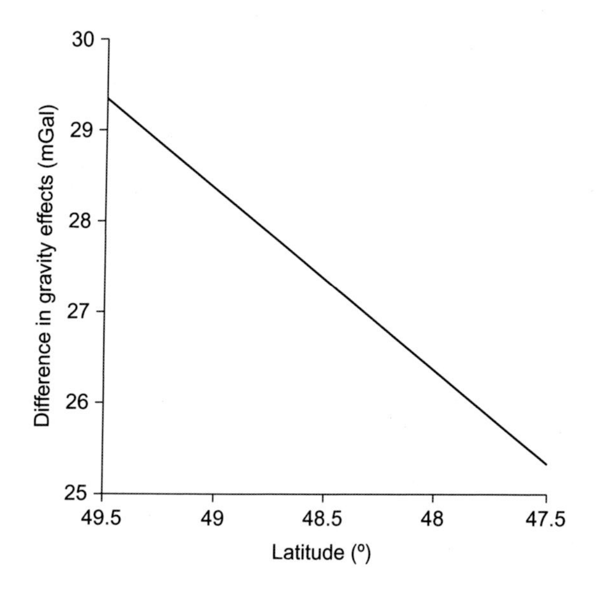

Figure 4.6 The difference between equipotential and layered solutions within the limiting latitudes of Slovakia.

Several pros and cons of both approaches were already mentioned above. The most important problem with the equipotential model is the ignorance of any possible density distribution within the real Earth, and the usage of the improper *GM* constant value what is a result of the unwarranted displacement of the atmospheric and especially the topographic masses. The advantage, on the other hand, is very simple and fast calculation and requires only four input parameters. The problem of the layered model is in the specific values of density in each layer, which come mostly from indirect measurements (they are prevailingly estimated from the velocities of seismic waves), and they are usually estimated for a spherical approximation of the Earth. From the other side, the advantages are the possibility of an arbitrary position of the calculation point or the generally realistic density distribution within the model.

4.6 RELATION OF THE NORMAL FIELD TO THE FREE-AIR CORRECTION

As it is commonly understood and as we already wrote above, the Bouguer anomaly is, roughly speaking, the measured gravity minus the expected gravity. This expected gravity is deemed to be caused by some "normal Earth" as its gravity effect. For our convenience, this effect is standardly composed of some partial effects which, within the process of the data processing, are usually applied as corrections to the measured gravity. The first effect or correction (and the largest one) is the normal field and the second is usually the free-air correction:

$$\Delta g_B = g_m - g_n + 0.3086h + \cdots \quad (4.9)$$

where Δg_B is the Bouguer anomaly, g_n is the normal field defined on the ellipsoid, $0.3086h$ is the free-air (or sometimes Faye's) correction, and h is the (unspecified) elevation. Please note that the literature is not in accord regarding the definition of the free-air correction. Some works consider that it contains also the normal field as such, while in other sources the free-air correction is understood solely as the $-0.3086h$ term. We have to mention here that certain part of the geophysical community still interpret the $-0.3086h$ term as the reduction (relocation) of the measured field to the surface of the ellipsoidal model. This approach, however, has been abandoned within the applied geophysics (LaFehr, 1991; Li and Götze, 2001; Hinze et al., 2005), and recently, Bouguer anomaly has been considered strictly as a measuring point or station quantity.

The examples of the discussed vertical gradient for the layered ellipsoidal model are listed in Table 4.3 along with the "standard" values (obtained from the second order formula for GRS80 according to Hinze et al. 2013, p. 133, Eq. 6.13 (Please note that there is a minus

Table 4.3 The Comparison of The Possible Free-Air Correction Values

Latitude (°)	GRS80 (mGal)		Layered ellipsoidal model (mGal)		Homogenous sphere (mGal)	
	$h=0$ m	$h=2000$ m	$h=0$ m	$h=2000$ m	$h=0$ m	$h=2000$ m
0	−0.308769	−0.308481	−0.308666	−0.308376	−0.308811	−0.308521
45	−0.308549	−0.308261	−0.308407	−0.308118	−0.308544	−0.308254
90	−0.308329	−0.308041	−0.308152	−0.307863	−0.308279	−0.307989

sign missing in front of the linear term in the cited source.) or NIMA, 2000, p. 4–2, Eq. 4.3). The gradient for homogenous sphere (with the "radius of equal volume") is presented for comparison, too.

If we rewrite Eq. (4.9) to

$$\Delta g_B = g_m - \left(g_n - 0.3086h\right) + \cdots \tag{4.10}$$

the whole term in the parentheses can be replaced either by the equation for the equipotential model in height h (Li and Götze, 2001) (and then to face the problem with the negative heights) or by the equation for the gravitational effect of the layered ellipsoidal model as described in this paper (any height). However, the latter will be more complicated.

4.7 CONCLUSION

We have tested the gravity effect of a layered ellipsoidal model as a possible alternative to the standard estimation of the so-called normal or theoretical gravity using the formula of Somigliana (1929). We also briefly comment some of the difficulties related to the two approaches. Our main goal however, was to initiate a broader discussion rather than to give any definitive solution.

We also briefly discuss the question of the free-air correction, which, in our approach, becomes an organic part of the normal gravity calculation and gets straightforward physical background.

While applied gravimetry does allow some approximations and has to work with the inherent ambiguity, it, on the other hand, does not allow any rearrangement of the Earth masses since that would be incompatible with the (geological) data interpretation. Although this may look rather peculiar, yet, it corresponds to the physical reality. From this aspect, the layered ellipsoidal model and its gravity effect could be an alternative way how to look at, and how to understand to the normal gravity despite the fact that there still are some important unanswered questions, both mentioned and unmentioned in our present contribution.

ACKNOWLEDGMENTS

This work was supported by the Slovak Research and Development Agency under the contracts APVV-0194-10 and APVV-0827-12. Special thanks are due to our colleagues Roman Pašteka, Pavol Zahorec and Juraj Papčo for their valuable discussions and for their continual interest in the topic in question. We also thank Dr. B. Meurers and second anonymous reviewer for their valuable comments.

REFERENCES

Bullen, K.E., 1975. The Earth's Density. Chapman & Hall, London.

Conway, J.T., 2000. Exact solutions for gravitational potential of a family of heterogeneous ellipsoids. Mon. Not. R. Astron. Soc. 316, 555–558.

Dyson, F.D., 1891. The potentials of ellipsoids of variable densities. Q. J. Pure Appl. Math. 25, 259–288.

Heiskanen, W.A., Moritz, H., 1967. Physical Geodesy. W. H. Freeman and Company.

Hinze, W.J., Aiken, C., Brozena, J., Coakley, B., Dater, D., Flanagan, G., et al., 2005. New standards for reducing gravity data: The North American gravity database. Geophysics 70, J25–J32.

Hinze, W.J., von Frese, R.R.B., Saad, A.H., 2013. Gravity and Magnetic Exploration. Cambridge University Press, 512 pp.

Jacoby, W., Smilde, P.L., 2009. Gravity Interpretation – Fundamentals and Application of Gravity Inversion and Geological Interpretation. Springer-Verlag, Berlin Heidelberg.

Kartvelishvili, K.M., 1982. Planetary Density Model and Normal Gravity Field of the Earth (in Russian). Publishing House "Nauka", Moscow.

LaFehr, T.R., 1991. Standardization in gravity reduction. Geophysics 56, 1170–1178.

Li, X., Götze, H.-J., 2001. Ellipsoid, geoid, gravity, geodesy and geophysics. Geophysics 66 (6), 1660–1668.

MacMillan, W.D., 1930. The Theory of The Potential. Dover Publications Inc, New York.

Mohr, P.J., Taylor, B.N., Newell, D.B., 2012. CODATA recommended values of the fundamental constants: 2010. Rev. Modern Phys. 84.

Moritz, H., 1968. Mass distributions for the equipotential ellipsoid. Bollettino di Geofisica Teorica ed Applicata 10 (37), 59–66.

Moritz, H., 1990, The Figure of the Earth: Wichman Verlag, Karlsruhe.

Nima Agency, 2000, TR 8350.2, Third Edition, Amendment 1, January 3, 2000, e-report.

Ochaba, Š., 1986, Geofyzika (in Slovak): SPN.

Petit, G., Luzun, B. (Eds.), 2010. IERS Conventions. IERS Conventions Centre, Frankfurt am Main.

Pizzeti, P., 1894. Sulla espressione della gravità alla superficie del geoide, supposto ellissoidic. Atti R. Accad. Lincei, Ser. V 3.

Rahman, M., 2001. On the Newtonian potentials of heterogeneous ellipsoids and elliptical disc. Proc. R. Soc. Lond. 457, 2227–2250.

Somigliana, C., 1929. Teoria generale del campo gravitazionale dell' ellissoide di rotazione. Mem. Soc. Astron. Ital. 4, 47 pp.

CHAPTER 5

Numerical Calculation of Terrain Correction Within the Bouguer Anomaly Evaluation (Program Toposk)

Pavol Zahorec[1], Ivan Marušiak[2], Ján Mikuška[2], Roman Pašteka[3] and Juraj Papčo[4]
[1]Slovak Academy of Sciences, Banská Bystrica, Slovak Republic [2]G-trend, s.r.o., Bratislava, Slovak Republic [3]Comenius University, Bratislava, Slovak Republic [4]Slovak University of Technology, Bratislava, Slovak Republic

5.1 INTRODUCTION

The new software Toposk was developed as part of the research project "Bouguer anomalies of new generation and the gravimetrical model of Western Carpathians" funded by the Slovak Research and Development Agency. One of the main tasks of this project was the unification and recalculation of the Slovak gravimetric database, which today contains approximately 320,000 points. Although our new program was designed primarily for the territory of Slovakia, it was later modified for a wider use, e.g., various coordinate systems and arbitrary subzone divisions were incorporated. There is also a possibility of using an arbitrary local orthogonal coordinate system for the inner zones calculation, which enables it for a global usage inside the continental areas. This version does not calculate bathymetric effects, as they relate to the territories outside the standard distance of 166.7 km in the case of inland countries like Slovakia. At the present time, we are developing a universal modification for worldwide calculations, which will also incorporate bathymetric corrections.

The program enables one to keep the real position of the calculated (measured) point with regard to the topographic surface, i.e., above, on or below it. It is relevant for example in the case of correcting the measured vertical gradient of gravity (VGG). There is also an option for using a density grid instead of a constant density of the topographic masses.

We have performed several numerical tests in order to check the quality of the computing algorithm. We have used simple synthetic

Understanding the Bouguer Anomaly. DOI: http://dx.doi.org/10.1016/B978-0-12-812913-5.00004-X

topography models represented e.g., by a cone, paraboloid, and others within these tests.

The software was programmed in C++ for 32 or 64 bit Windows applications.

5.2 MAIN FEATURES OF THE NEW SOFTWARE TOPOSK

The main idea is the straightforward calculation of the gravitational effect of the topographic masses, which we call the topographic effect (sometimes, it is called the "mass correction," e.g., Hammer, 1974 or Meurers et al., 2001b). The terrain correction is then derived from the topographic effect by its subtraction from the gravitational effect of the truncated spherical layer (or vertical cylinder in the planar approach within the inner zones). The relationship between the masses considered in terrain correction and topographic effect evaluation, respectively, is clearly demonstrated in Fig. 5.1.

We perform the calculation up to the standard angular distance 1°29′58″ (approximately 166.7 km), which is the outer limit of zone O of the Hayford–Bowie system (Hayford and Bowie, 1912). The dividing of this calculated area into several zones follows the traditional approach originated in former Czechoslovakia, where Pick et al. (1960) defined inner or local zone up to 5.24 km (the outer limit of the zone H of the Hayford–Bowie system), which was later (Bližkovský et al., 1976) divided into the zones T1 (square 500 × 500 m; today, we use circular zone with radius 250 m instead of that square area) and T2 (up to 5240 m). Outer zone was later modified to the zones T31 (5240–28,800 m) and T32 (28.8–166.7 km) after Mikuška and Grand (1989). The radius 28,800 m corresponds to the outer limit of the zone L of the Hayford–Bowie system. Different digital elevation models (DEMs), with increasing resolution toward the calculation point, are used within particular zones. This system (Fig. 5.2) was proven as a

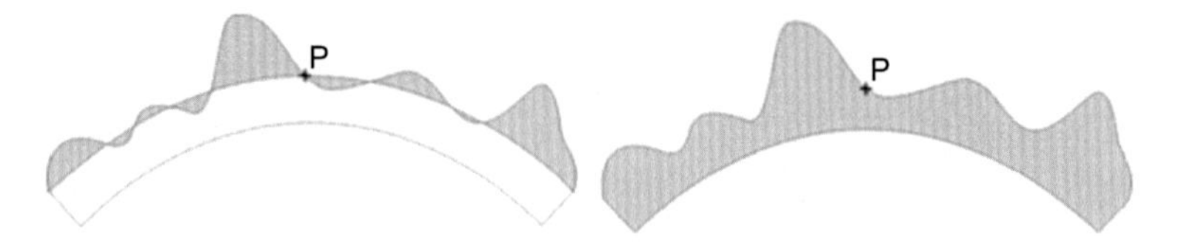

Figure 5.1 Relation between the masses considered in terrain correction (left) and topographic effect (right) calculation at point P.

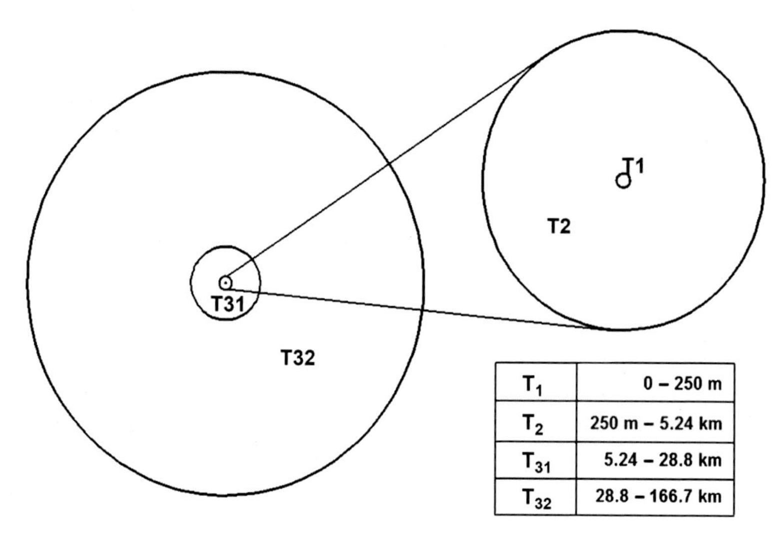

Figure 5.2 Standard system of terrain correction zone divisions used in Slovakia, after Grand et al. (2004).

suitable approach for the terrain corrections recalculation of the Slovak regional gravity database (Grand et al., 2001, 2004), except the inner zone T1, where the DEM used in that work was insufficient. Therefore, we have focused our attention particularly on this zone regarding the more detailed elevation models, as well as the deeper study of the concept of the interpolated heights of the calculation points (Zahorec, 2015). The inner (T1) and intermediate (T2) zones are treated in planar approach (this yields a small, in most cases, negligible error), and DEMs in local orthogonal coordinates are used. Outer zones are treated using a spherical approach; the calculation is based on the DEMs in ellipsoidal coordinates (e.g., ETRS89). The program enables the transformation between known local and global coordinate systems used in the area of Slovakia (Slovak national systems JTSK, S-1942, global systems UTM 33/34 and ETRS89/WGS84). We use a system of normal heights—Kronstad base system Baltic after adjustment (Bpv). Transformation between direct measured ellipsoidal heights from GNSS (in ETRS89, ellipsoid GRS80) to this local height system Bpv is made using Slovak local quasigeoid DVRM (Klobušiak et al., 2005) related to the same ellipsoid GRS80. The mentioned subdivision and the limits of the individual zones can be changed by the user.

The calculation can be performed utilizing a set of scattered points of known elevation (acquired, e.g., by in situ geodetic measurements; e.g., Lyman et al., 1997; LaFehr et al., 1988) instead of a standard DEM grid, using a set of triangular facets created by means of the triangulation method of Joe (1991). This option was introduced because of the possible imperfection of gridding processes in the case of irregularly distributed elevation (measured) data. However, this approach is more time consuming than the standard calculation with a DEM grid. There is a check box *Or use toposk database* just under the selection box *Elevation grid* in the T1 and T2 tabsheet for this option (Fig. 5.6).

Although various numerical approaches can be used for the calculation of the topographic effect within particular zones, the formula of Pohánka (1988) for a 3D polyhedral body is preferred. This formula makes the calculation of the topographic effect possible in an arbitrary point, so also inside the topographic masses. This property is very useful for the calculation of the effect of topography related to VGG measurements, underground gravity measurements, and others. It enables us also to calculate the topographic effect exactly in the gravity meter sensor position, which is required for the case of precise geodetic absolute gravity measurements.

The calculation is performed by default for a given constant density of the topographic masses (e.g., 2.67 g/cm^3). But there is also an option to make the calculation for variable densities defined by a density grid, if it is available. In this case, the topographical effect of each segment (prism) within the particular zones is calculated using the supplied density. The total computational time for all zones in today's common PC (CPU 3.4 GHz) is less than one second per point.

5.3 INNER ZONE T1

Our standard radius for the inner zone T1 is 250 m. The topography within this zone is approximated by one 3D polyhedral body (see an example in Fig. 5.3), the gravitational effect of which is calculated using the formula of Pohánka (1988). The upper surface of the polyhedral body is created by a triangular net constructed from the elevation data, which are usually in a grid format. The transition between the rectangular grid and the circular boundary of the zone is realized by

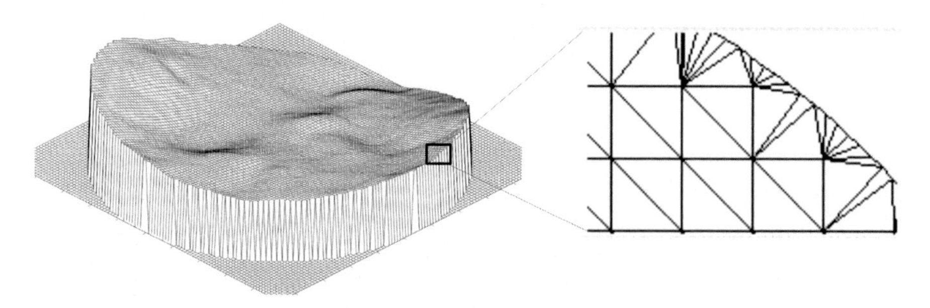

Figure 5.3 Example of 3D polyhedral body approximation of the topography within the inner zone T1 and a detailed sketch of the transition area along its boundary.

additional 360 points interpolated from the grid along the zone boundary (Fig. 5.3). The interpolation method is at choice (see next section), a bilinear interpolation as default.

The terrain correction is obtained as the difference between the gravitational effect of the vertical cylinder (with the same radius and the height equal to the calculation point elevation) and the calculated topographic effect. During the recalculation of the regional gravimetrical database, we used a high resolution DEM with grid cell size of 10 m (DMR-3, Topographic Institute, 2012), which is considerably better than the previous models. However, detailed or microgravity surveys need even greater detail for the DEM (perhaps 1×1 m, which typically can only be obtained from mentioned in-field surveying).

The interpolated heights of the calculation points can be used instead of the measured heights for the inner zone terrain corrections. This approach reduces the errors resulting from the discordance between the real (measured) and the model heights of the calculation points. Those height differences can achieve 100 m or even more in mountainous areas, which leads to errors of several mGal within terrain corrections for zone T1. We have made a detailed study of the optimal distance for use of interpolated heights. The tests with real data as well as synthetic topography models suggest that the optimal distance depends on the quality of the DEM. In most cases, it could be less than 250 m (Zahorec, 2015). Anyway, by using interpolated heights of calculation points to a distance of 250 m, we observe lower errors than when using measured heights.

5.4 INTERMEDIATE ZONE T2

We use the standard extent for this zone 250–5240 m from the point of calculation. The topography within this zone can be approximated by two methods: by a set of triangular prisms or by a set of segments of a vertical cylinder. The first method is the analogy of the polyhedral body method applied within the inner zone T1, where the gravitational effect of the prism is calculated by Pohánka's formula. For the second method, we can designate using the classic or template approach, as can be seen in Fig. 5.4. Both methods give similar results in the case of detailed DEM but the classic method is considerably faster.

The heights of particular segments (prisms) are estimated by one of the standard interpolation methods, e.g., bilinear interpolation, average, bicubic spline, and others. We leave the choice of interpolation method to user experience and standardly use a bilinear interpolation because it is simple and sufficient in the case of detailed elevation grids. The segment size should be approximately the same as the grid cell size of the DEM. Today, we typically use elevation grids with cell sizes of 30×30 m. The terrain correction T2 is obtained as the difference between the gravitational effect of the hollow vertical cylinder (with the corresponding inner and outer radius and with height equal

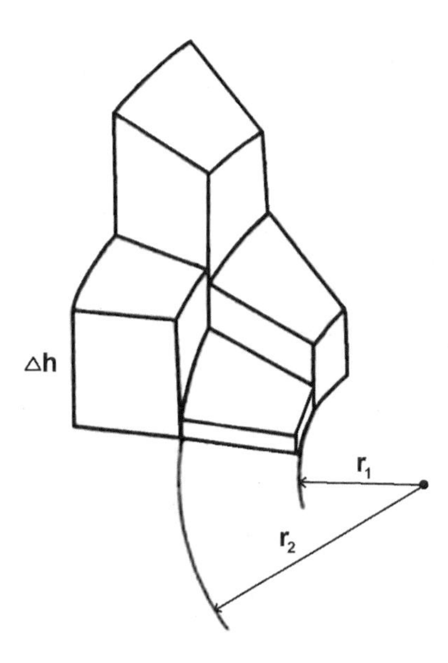

Figure 5.4 "Classic" approximation of topography using vertical cylinder segments, r1 and r2 are inner and outer radii of the particular segment, Δh is the mean height of the segment

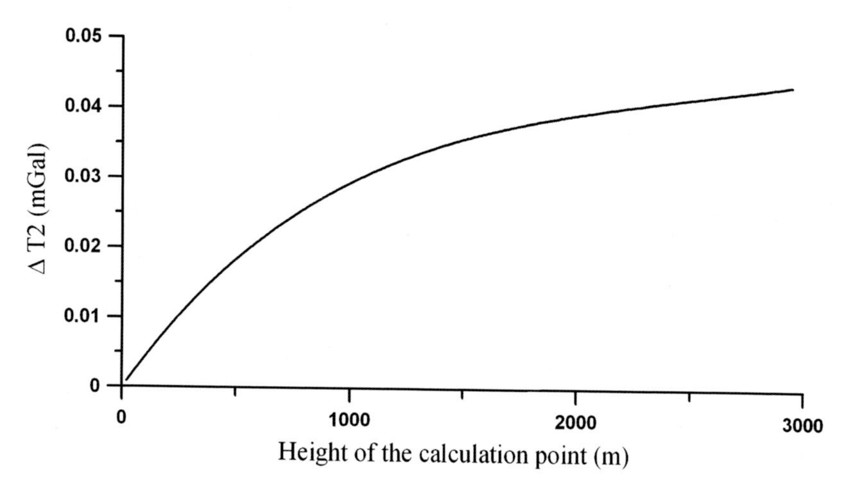

Figure 5.5 Difference between the spherical and planar concepts of the terrain corrections T2 (2.67 g/cm^3).

to the calculation point elevation) and the calculated topographic effect. We have estimated the error resulting from the planar approximation within the intermediate zone. We compared the planar and spherical approaches to the calculation of the terrain correction T2 on a set of synthetic calculation points covering the elevation range that occurs in Slovakia. An elevation model with constant zero level was used within this test to obtain maximal T2 values. The maximum expected error is in the range of a few tens of μGal for a density 2.67 g/cm^3, see Fig. 5.5.

5.5 OUTER ZONES T31 AND T32

The standard distances which we use for these zones are 5240–28,800 m and 28,880–166,730 m, respectively. The topography can also be approximated by two methods: by the set of triangular prisms, this is similar to the zone T2 but here in a spherical modification (geographic coordinates are transformed to Cartesian, and the position and shape of prisms is adapted to the sphere). The second method is represented by the set of segments of the spherical layer, where their effects are calculated by the formula of Mikuška et al. (2006). Both methods give virtually the same results, but the latter method is faster. The elevation grids based on ellipsoidal coordinates WGS84 or ETRS89 are used, e.g., from SRTM data (Jarvis et al., 2008). Today, we use grids $3 \times 3''$ for T31 and $30 \times 30''$ for T32 calculations. The terrain correction within these zones is obtained as the difference between the gravitational effect of the truncated spherical layer (with the corresponding inner and outer radii and the height equal to the calculation point elevation) and the sum of the gravitational effects of all segments in the given zone.

The calculation of the gravitational effect of each segment is made for a given constant density, or for a density interpolated from the density model (grid). Such option is available for all zones.

5.6 USER INTERFACE

The user interface is shown in Fig. 5.6. The upper section is designed for the choice of input file, coordinates type, and the density. The program enables to check the correctness and range of known coordinates types.

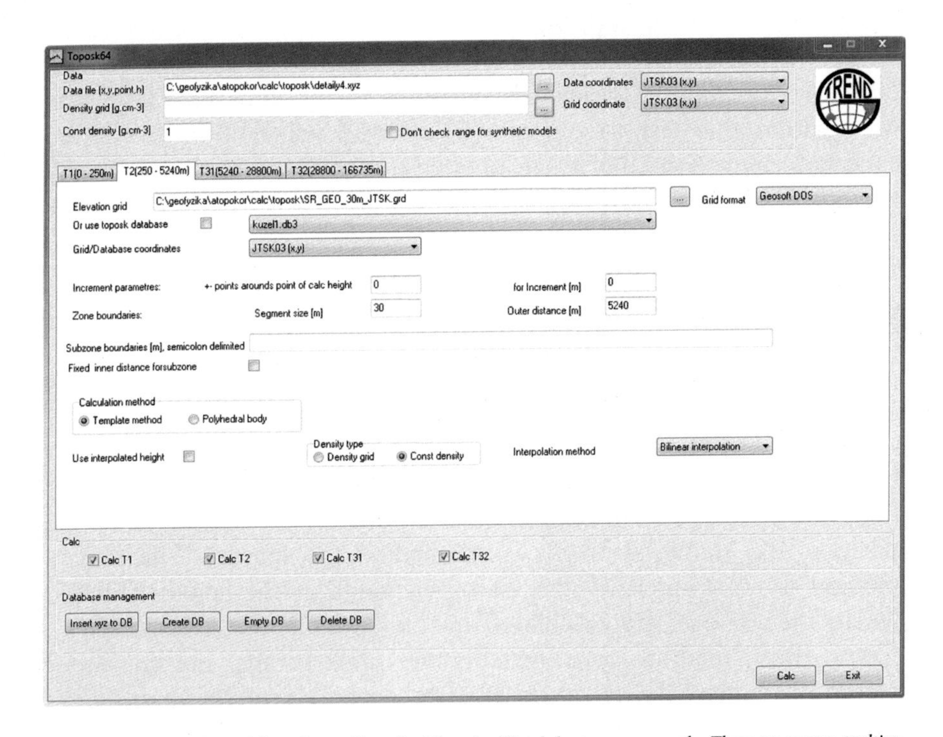

Figure 5.6 User interface of the software Toposk with active T2 tabsheet as an example. There are pop-up tooltips connected to all check boxes.

```
--------------------------------------------------------------------------
Data name.............: C:\geofyzika\atopokor\calc\toposk\database_points.xyz
Data coordinate.......: JTSK03 (x,y)
Increment [m].........: 0.000000
Use interpolate height: NO
Grid name.............: C:\geofyzika\atopokor\calc\toposk\SR_GEO_30m_JTSK.grd
Grid coordinate.......: JTSK03
Grid type.............: Surfer 6 binary
Interpolation method..: Bilinear interpolation
Inner distance [m]....: 250.000000
Outer distance [m]....: 5240.000000
Calculation method....: Template method
--------------------------------------------------------------------------
         X           Y       point        H         H-Hi      NTE2         T2
 -372082.5  -1181817.9        501   1353.179       3.122   35.98968   4.01673
 -368073.6  -1187666.5        502    956.366       2.416   25.93893   1.39551
 -340764.5  -1186926.9        503   1675.012      -1.325   45.89653   3.67710
 -346379.2  -1186669.2        504   1505.837      -0.613   40.93490   3.69174
 -359324.1  -1187170.4        505    929.189       1.276   25.42678   1.00804
 -335894.6  -1186792.1        506   1306.620      -3.726   35.96564   2.60247
 -344292.3  -1177741.4        507   1026.553      -1.276   26.72140   2.91932
 -332727.1  -1174225.0        508    919.482       1.200   25.52640   0.58624
```

Figure 5.7 Example of output file from the T2 zone calculation.

The main section contains four tabsheets, which control the parameters of the calculation for each zone: the elevation model, the zone boundaries, the segment size, the calculation method, the interpolation method, and others. In the bottom section, the user can manage the elevation-data databases (SQLite system) in the case of nongrid calculation.

The input data file must contain the coordinates and elevations of the calculation points. An example of the output file for the zone T2 calculation is shown in Fig. 5.7. There are the calculated values of topographic effect (NTE2—"near topographic effect") and terrain correction T2 for the given density (in mGal), as well as the parameter $H-H_i$ for each point, which is the difference between measured elevation and the elevation interpolated from the DEM. This parameter gives useful information about the quality of the elevation data (DEM) used for the respective terrain correction calculation.

5.7 PROGRAM TESTING ON SYNTHETIC DATA

We have performed several tests to check the quality of the numerical algorithms implemented in the Toposk approach. We focused on simple synthetic models of the topography, the analytical gravitational effects of which are possible to calculate by means of closed formulas (the cone, paraboloid, planar, or spherical layer) and compare them

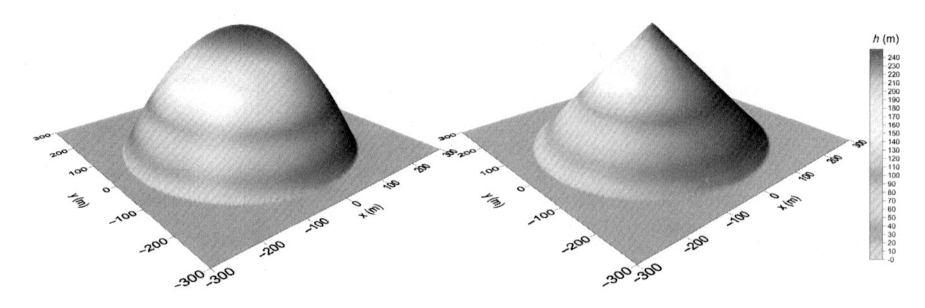

Figure 5.8 Simple topography models approximated by the paraboloid and cone.

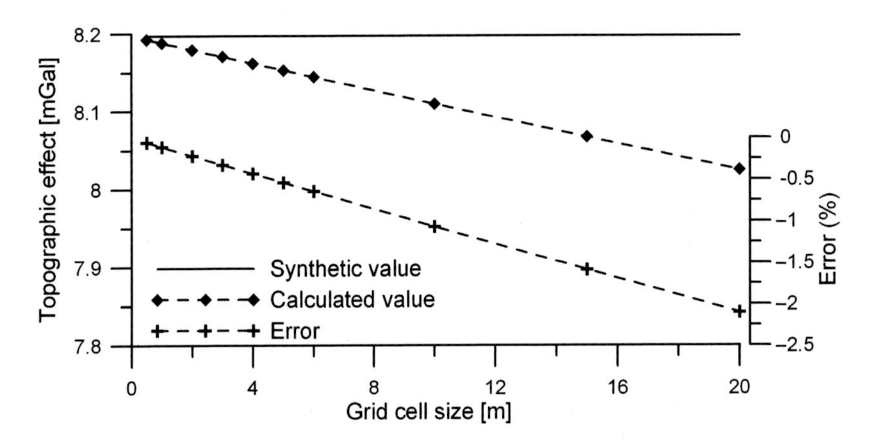

Figure 5.9 Test on conus-like topography showing the calculation error dependence on the model resolution. The calculation point is situated on the cone apex.

with the calculated topographic effects, see Fig. 5.8. We have used common formulas found in the literature for the analytical calculations of the gravitational effect at the cone (paraboloid) apex (e.g., Helmert, 1884; Hammer, 1939; Válek, 1969).

The tests have confirmed that the calculation error is purely a function of the resolution of the model (grid). Highly detailed models (e.g., grid cell size 1×1 m) lead to errors only of few μGal, see Fig. 5.9. The important question was to confirm the accuracy of the calculation in the case when the calculation point is above or below the surface (i.e., inside the topographic masses, e.g., in the case of underground gravity measurements). Programs previously in use in Slovakia did not allow such situations. In Fig. 5.10, we show the results of a comparison of analytical and calculated gravitational (topographic) effect along the

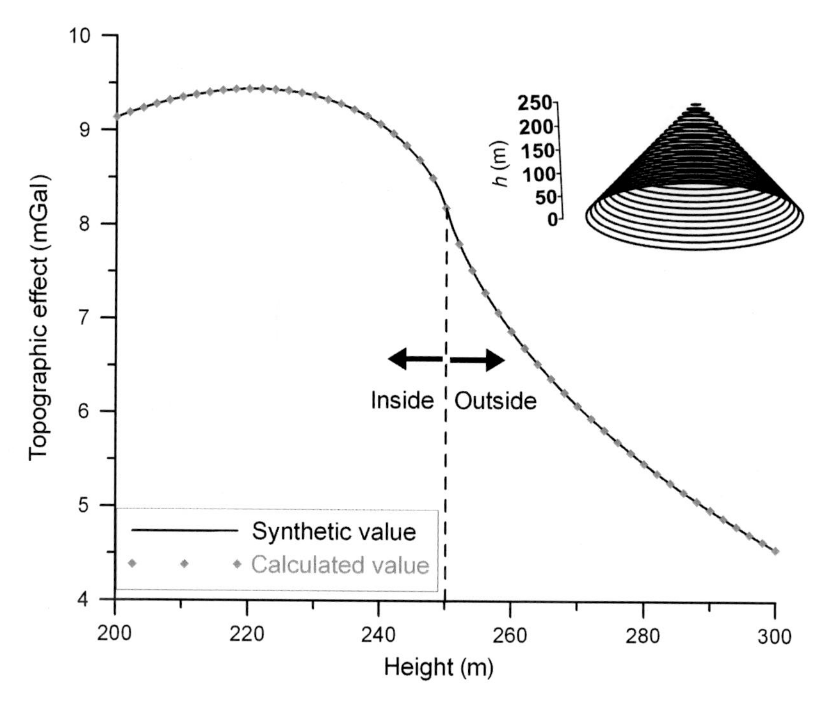

Figure 5.10 Comparison of the analytical value and program Toposk-calculated effects of a cone along its axis. The dashed line shows the calculation point with height of 250 m, lying directly on the cone apex.

cone axis. The analytical values were calculated by the formula published in Meurers (2001a). The results confirm the perfect coincidence of the calculated values in the case of very detailed topography model (e.g., 1×1 m).

5.8 REAL DATA CALCULATIONS

We have proven the program by calculations also on the basis of real measured data. This is particularly important for measurements of VGG (or rather tower VGG, as we determine it by means of the gravity measurements at different height levels) because of the sensitivity to near-station topography (e.g., Zahorec et al., 2014). The measured values of tower VGG can deviate considerably from the "expected" or "normal" value (-0.3086 mGal/m) especially in mountainous regions, as is shown in Fig. 5.11 (black symbols). After allowing for the topographic effect, the corrected values (blue crosses in Fig. 5.11) are considerably closer to the normal gradient value.

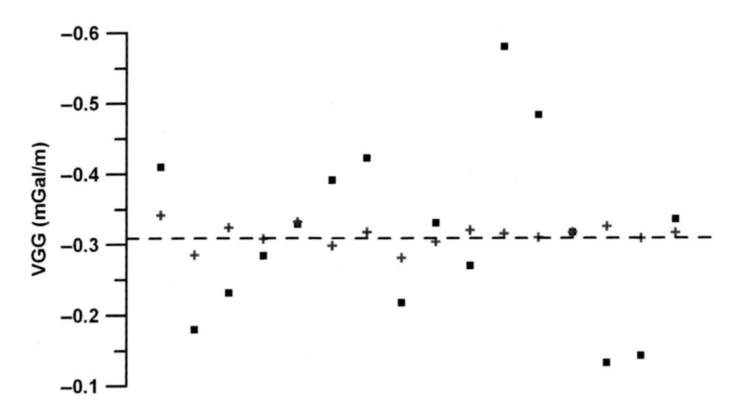

Figure 5.11 Set of field VGG measurements in Slovakia. The black squares represent measured values (Min = − 0.580, Max = − 0.132 mGal/m), whereas the blue (gray in print versions) crosses represent values after the correction for topography effect (Min = − 0.342, Max = − 0.281 mGal/m). Dashed line represents the normal VGG (−0.3086 mGal/m).

5.9 CONCLUSIONS

As the most important attribute of the new software Toposk, we have shown the correct calculation of the topographic effect and terrain corrections for the arbitrary point location with regard to the real Earth surface (above or below it). The detailed approximation of topography within the inner zone by 3D polyhedral body has been applied. Alternative computing methods, namely the "classic" segment-based methods and the vertical prism method, are available within the intermediate and the outer zones. Several tests on synthetic topography models proved the correctness of the implemented algorithms. The optional concept of using interpolated heights for the calculation points within the nearest zone is recommended. We have also estimated the maximum expected error resulting from use of a planar approach within zone T2 (up to 5240 m) to a few tens of μGal (for density 2.67 g/cm^3). The user has a choice to either use the default zone division (to four zones with radii 250, 5240, 28,800, and 166,730 m) or define custom zones. The possibility of using various local coordinate systems leads to more universal usage of the software. The use of scattered elevation data instead of an elevation grid is also available, however, this option is more time consuming.

This software was recently used during the recalculation of the complete Slovak gravimetrical database (see Chapter 7, National gravimetric database of the Slovak Republic in this book). The concept of interpolated heights of calculation points was applied within the nearest zone T1 (0–250 m).

ACKNOWLEDGMENTS

The authors are grateful to the Slovak Research and Development Agency APVV (grants No. APVV-0194-10, APVV-0827-12, and APVV-0724-11) and the Slovak Grant Agency VEGA (grants No. 2/0067/12 and 1/0095/12) for the support of the software development. We thank to Allen Cogbill, John Bain, and one anonymous reviewer for improving the manuscript.

REFERENCES

Bližkovský, M., Bednář, J., Klečka, K., Odstrčil, J., 1976. Instruction for Gravimetrical Mapping in the Scale 1: 25 000. MS, Czech Geological Authority, Prague (in Czech).

Grand, T., Šefara, J., Pašteka, R., Bielik, M., Daniel, S., 2001. Atlas of Geophysical Maps and Profiles. Part D1: Gravimetry. Final Report. State Geological Institute, Bratislava, MS Geofond (in Slovak).

Grand, T., Pašteka, R., Šefara, J., 2004. New version of terrain correction in the Slovak regional gravity database. Contrib. Geophys. Geodesy 34, 315–337.

Hammer, S., 1939. Terrain corrections for gravimeter stations. Geophysics 4, 184–194.

Hammer, S., 1974. Topographic and terrain correction for airborne gravity. Geophysics 39, 537–542.

Hayford, J.F., Bowie, W., 1912. The effect of topography and isostatic compensation upon the intensity of gravity. U.S. Coast and Geodetic Survey, Washington, DC, Special Publication No. 10.

Helmert, F.R., 1884. Die mathematischen und physikalischen Theorieen der höheren Geodäsie. II. Teil, Teubner, Leipzig.

Jarvis, A., Reuter, H.I., Nelson, A., Guevara, E., 2008. Hole-filled SRTM for the Globe Version 4. Available from the CGIAR-CSI SRTM 90 m Database: http://srtm.csi.cgiar.org.

Joe, B., 1991. GEOMPACK – a software package for the generation of meshes using geometric algorithms. Adv. Eng. Softw. 13, 325–331.

Klobušiak, M., Leitmanová, K., Ferianc, D., 2005. Realization of obligatory transformation between national coordinates and height reference system into ETRS89. Proceedings of the International Conference Tatry 2005 (in Slovak).

LaFehr, T.R., Yarger, H.L., Bain, J.E., 1988. Comprehensive treatment of terrain corrections with examples from Sheep Mountain, Wyoming. 58th Ann. Internat. Mtg., Sot. Explor. Geophys., expanded Abstracts. 361–363.

Lyman, G.D., Aiken, C.L., Cogbill, A., Balde, M., Lide, C., 1997. Terrain mapping by reflectorless laser range finding systems for inner zone gravity terrain corrections. Expanded Abstracts, 1997 SEG annual meeting, November 2–7, Dallas, TX.

Meurers, B., 2001a. Remarks on the discontinuity of the gravity gradient at the apex of a cone. Proceedings of the 8th International meeting on Alpine gravimetry, Leoben 2000, Oesterr. Beitraege zu Meteorologie und Geophysik, Heft 26, pp. 181–186.

Meurers, B., Ruess, D., Graf, J., 2001b. A program system for high precise Bouguer gravity detemination. Proceedings of the 8th International meeting on Alpine gravimetry, Leoben 2000, Oesterr. Beitraege zu Meteorologie und Geophysik, Heft 26, pp. 217–226.

Mikuška, J., Grand, T., 1989: Calculation of topographic corrections T3 by means of 8-bit computer Sinclair ZX Spectrum. Manuscript, Geofyzika Bratislava (in Slovak).

Mikuška, J., Pašteka, R., Marušiak, I., 2006. Estimation of distant relief effect in gravimetry. Geophysics 71, J59–J69.

Píck, M., Pícha, J., Vyskočil, V., 1960. Gravity topographic corrections for the territory of Czechoslovakia, Travaux Géophysiques, 129. pp. 113–129.

Pohánka, V., 1988. Optimum expression for computation of the gravity field of a homogenous polyhedral body. Geophys. Prospect. 36, 733–751.

Topographic Institute, 2012. Digital Terrain Model Version 3 (Online). http://www.topu.mil.sk/14971/digitalny-model-reliefu-urovne-3-%28dmr-3%29.php.

Válek, R., 1969. Gravimetry III – Direct and Inverse Problem, Earth Gravity Field and its Anomalies. SPN, Praha, in Czech.

Zahorec, P., 2015. Inner zone terrain correction calculation using interpolated heights. Contribut. Geophys. Geodesy 45/3, 219–235.

Zahorec, P., Papčo, J., Mikolaj, M., Pašteka, R., Szalaiová, V., 2014. The role of near topography and building effects in vertical gravity gradients approximation. First Break Vol. 32/1, 65–71.

CHAPTER 6

Efficient Mass Correction Using an Adaptive Method

Wolfgang Szwillus and Hans-Jürgen Götze
Christian Albrechts University, Kiel, Germany

6.1 INTRODUCTION

The gravity effect of topography veils the gravity signal coming from deeper sources. The process of removing the gravity effect of topography is called "mass correction." Although the necessity to correct for the influence of topography on gravity has been recognized already in the 19th century, progress in this area continues. For instance, Mikuška et al. (2006) have challenged the convention, that only topography up to a distance of 167 km from the station is relevant (see e.g., Bullard, 1936; LaFehr, 1991). They proposed to calculate the "distant relief effect" (DRE), which refers to the gravity of topography beyond a correction radius of 167 km, and found long wavelength, yet significant contribution of distant relief.

The DRE is less important for density modeling at a local exploration scale because long wavelength trends are removed during regional–residual separation (e.g., Nabighian et al., 2005). However, for modeling deep and/or extensive structures, accurate removal of long wavelength effects is critical. These wavelengths are now provided with unprecedented accuracy by satellite measurements (Bouman et al., 2015).

Therefore, a new algorithm for topographic correction should be able to calculate the mass correction from global topography or at least from large areas. Furthermore, it must correct ground, airborne, and satellite measurements consistently. Thus, it must also be capable of handling high-resolution topographic data efficiently.

Mass correction can be calculated directly, or in two steps (giving the same result). In the two-step procedure, first a "Bouguer correction" is applied, which assumes topography is an infinite flat slab, with

Understanding the Bouguer Anomaly. DOI: http://dx.doi.org/10.1016/B978-0-12-812913-5.00005-1

thickness equal to the station height. Afterward, a "topographical correction" is carried out, which corrects for the difference between real topography and the assumed slab. The adaptive method (Szwillus, 2014) we propose works with either convention, but we prefer the one-step procedure because it is conceptually simpler.

All methods for topographic correction based on digital elevation models (DEMs) work essentially the same way: Each grid cell of the DEM is translated into an elementary volume, for example, into a flat-top prism. Repeating this for all grid cells gives a volumetric model of topography. One of the earliest papers which describe the application of DEMs for gravity-field corrections was the one of Cogbill (1990). To calculate the gravity effect of such a volume model, topographic and bathymetric densities are needed. Typically, constant densities of 2670 and 1630 kg/m^3, respectively, are assumed. Existing methods make use of different elementary volumes e.g., cylinder segments (Hayford and Bowie, 1912; Hammer, 1939), prisms (Nagy et al., 2000), or tesseroids (Grombein et al., 2013). Generally, there exist several ways to calculate the gravity effect of a volume: closed analytical solutions, analytical estimation by Taylor series, or numerical integration (Heck and Seitz, 2007).

For any method, the number of calculations is proportional to the number of pairs of grid-points and stations. Accordingly, if the topographical grid contains $X \times Y$ cells, and there are N stations, a total of $X \times Y \times N$ pairs need to be calculated. Increasing the resolution or horizontal grid extent, and therefore, the number of nodes in the grid makes the calculations a lot more time consuming.

To reduce the processing time, we propose an adaptive approach that uses the gravity field itself as a way to determine the necessary resolution. Starting with a coarse grid for representation of topography, the resolution is increased locally, until a further increase of resolution would only change the obtained topographic effect negligibly.

The aim of this work is to demonstrate the advantage of the adaptive approach over earlier applications. It needs only a fraction of the computational time and still results in sufficiently accurate value for gravity-field corrections.

6.2 PRINCIPLE OF ADAPTIVE ALGORITHM

Hayford and Bowie (1912) already used a variable resolution representation of topography over hundred years ago. The cylinder compartments are getting larger as one moves away from the gravity station due to the $1/r^2$ numerical effect of the gravity field decay. We build on this approach, but change the size of the elementary bodies (i.e., the resolution) automatically, instead of defining an a priori size. In essence, the calculation of the gravity effect of many small elementary bodies is replaced by calculating fewer but partly larger blocks.

6.2.1 Variable resolution representation of topography

Quadtrees (Samet, 1990) provide an elegant way to create variable resolution representations of a grid. Here, a topographic grid, called the base grid, is used as input. For simplicity's sake, we will assume the base grid is square and contains $2n$ values in each dimension. Then the associated quadtree consists of $n + 1$ layers. Each layer contains nodes that correspond to rectangular cells (Fig. 6.1 and 6.2). Every node stores the topographical height and density of its corresponding cell.

The nodes of the different layers are in a hierarchical relation. This is achieved by constructing the quadtree from bottom up. The cells of layer $j + 1$ are created by merging four neighboring cells of layer j. Conversely, the cells of layer j can be obtained by splitting the cells of layer $j + 1$. Let N be a node of layer $j + 1$, then four nodes from layer j exist, that were used to construct node N (see Fig. 6.1). These four nodes C1, C2, C3, and C4 are the children of N. Similarly, N is called the parent of C1 to C4. The top layer (layer n) contains only one node, called the root (note that the root is at the top of the tree!). Its density

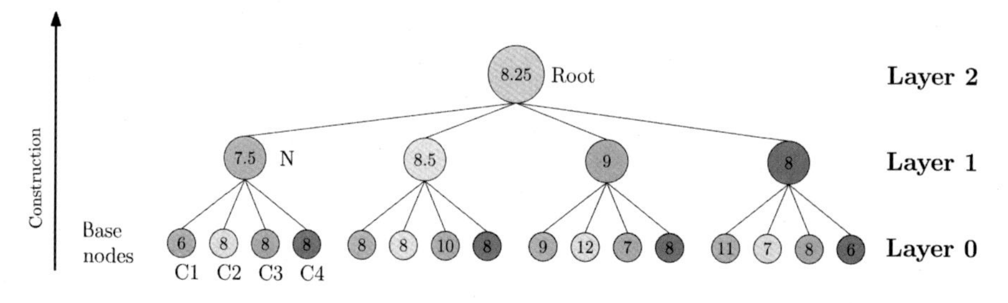

Figure 6.1 Tree representation of the topography. Colors are used to mark corresponding grid cells in Fig. 6.2. The values of the nodes in layer 1 are obtained by averaging over four nodes of layer 0 (cells of the base grid). Likewise is the final calculation for the single node in layer 2, which is the root of this tree. Thus, the tree is constructed from bottom up. Note that the root is at the top of tree.

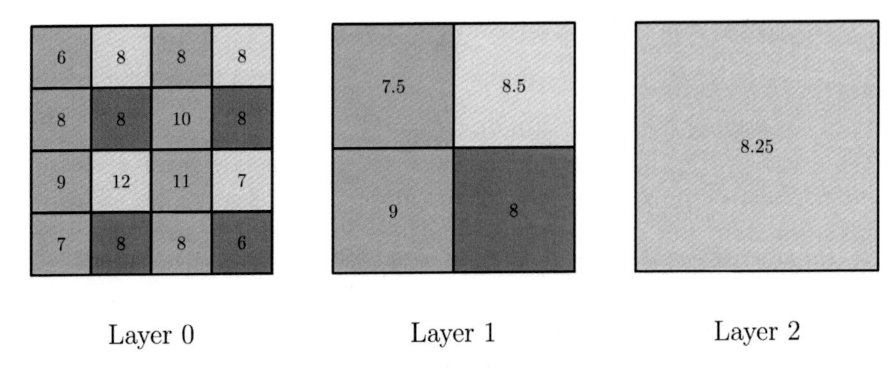

Figure 6.2 Grid representation of topography at different resolutions. The cells of layer 1 are obtained by averaging over four cells of layer 0, and likewise for layer 2 (gray in print versions).

and height are equal to the average density and height of the entire region.

Each layer has half the resolution of the layer below it. Thus, a quadtree essentially represents a set of n grids at different resolution steps. These grids are nested in such a way that the resolution can be changed locally without affecting other parts of the grid.

The heights and densities stored in the nodes of layer $j+1$ have to be computed by averaging four nodes of layer j. The topographical mass belonging to a single node is approximately:

$$m_i = A_i h_i \rho_i$$

where A_i is the area, h_i is the height, and ρ_i is the density of each children cell i ($i = 1.4$). Area, height, and density of the parent cell ($\overline{A}$, $\overline{h}$, and $\overline{\rho}$) must be determined from the children in such a way that total mass topographic mass is conserved. This guarantees an optimal approximation of the gravity effect by the higher layers.

The area of the parent cell is just the sum of the children's areas:

$$\overline{A} = \sum_i A_i$$

We ensure mass conservation by first calculating the density of the parent cell, using area-weighted averaging:

$$\overline{\rho} = \frac{\sum_i A_i \rho_i}{\sum_i A_i} = \frac{\sum_i A_i \rho_i}{\overline{A}}$$

Then, we calculate the height of the parent cell by averaging height weighted with area and density:

$$\overline{h} = \frac{\sum_i A_i \rho_i h_i}{\sum_i A_i \rho_i} = \frac{\sum_i A_i \rho_i h_i}{\overline{A}\overline{\rho}}$$

This scheme of averaging is a stable procedure because the divisor in the above equation is always nonzero, at least if all children densities are larger than zero.

6.2.2 Adaptive Algorithm

The adaptive algorithm calculates the gravity effect caused by a topographic grid on a set of gravity stations. If surface density data are available, they can be included as well. The gravity value returned by the algorithm is only an approximation of the "true" gravity value, which is defined as the gravity effect of layer 0 (i.e., in case of nonadaptive calculations at original resolution). The maximum acceptable deviation between true gravity value and the result of the algorithm is an input parameter called the *error tolerance*.

The second parameter of the adaptive algorithm is the "starting layer" j^*. It defines a maximum cell size and respectively a minimum resolution. The adaptive algorithm then refines the grid locally starting from this cell size. In principle, the refinement could start a layer n. Then, the entire grid is represented as a single block. However, when doing so, inaccurate results may be obtained (see tests in the next section). Setting an adequate maximum cell size stabilizes the algorithm.

Our algorithm is based on classic adaptive algorithms for numerical integration (e.g., Gander and Gautschi, 2000). Their central idea is to compute two estimates of an integral over a domain. One uses a coarse resolution, and another uses a finer resolution. Their difference provides a measure of the uncertainty/error of the estimates. If this estimate surpasses an error threshold, the domain is split, and the algorithm is applied recursively to all subdomains. This approach can be applied to the calculation of the mass correction, since it is essentially a form of integration as well.

The adaptive algorithm traverses the tree from top to bottom, beginning with the nodes of the starting layer j^*. For each node N in layer j^*:

1. Compute two gravity values: That of the node itself, and the total gravity of its four children.
2. Calculate the difference of those gravity values to get an estimate of the uncertainty of the gravity.
   a. If the uncertainty is higher than the layer error threshold (note: this is not the global error tolerance) try to split each children of N into their four children (the "grand-children" of N).
      i. If this is possible, recursively repeat steps 1 and 2 for all children of node N.
      ii. If node N has no grandchildren, the children of N are already from layer 0, and thus cannot be split. Save the gravity value and proceed to the next node of the starting layer j^*.
   b. If the uncertainty is below the local error threshold, save the gravity value and proceed to the next node of the starting layer j^*.

The layer error threshold is equal to the global error tolerance divided by the number of nodes each active layer. This procedure is repeated for each gravity station. All stations are processed completely independently. Stations can be located anywhere above the topography.

6.2.3 Forward Calculation of Gravity

The adaptive algorithm needs a way to compute the gravity value of a node in the quadtree. Since we assume a constant topographic height in each grid cell, each cell automatically corresponds to a tesseroid ("spherical prism"), because global DEMs are in geographic coordinates. There is no closed analytical expression for the gravity field of a tesseroid (Heck and Seitz, 2007), because the corresponding Newton-Integral is elliptical. However, the gravity effect of a tesseroid can be estimated using numerical methods (Grombein et al., 2013). Here, we use the adaptive Gauss–Legendre integration for a surface integral formulation of the Newton-Integral. Adaptive numeric integration allows calculating the gravity effect for each tesseroid to a specified accuracy.

Note that the adaptive approach itself is agnostic with respect to the elementary bodies. Replacing tesseroids by prisms, for example, would be straightforward. The only requirement is that the forward calculation for each elementary body can be determined accurately enough. Specifically, the accuracy of the forward calculation needs to be less than the smallest layer error tolerance. Otherwise, the errors of the forward gravity calculation would dominate, making error estimation (see step 2 of the algorithm) meaningless. Holzrichter (2013) has developed a similar adaptive algorithm that uses a polyhedral approximation of topography.

6.3 TEST OF ALGORITHM

The acceptance of the new approach depends on reliability and efficiency of the adaptive algorithm. Reliability means the true error does not exceed the error tolerance, while efficiency points to faster calculations—faster than calculations without adaptive change of resolution. Furthermore, the impact of the parameters (error tolerance and minimum resolution) on the results of the algorithm needs to be investigated. As a normal procedure, the new adaptive algorithm is tested by a "two-level" approach: the first bases on numerical experiments with synthetic topography and the second on a global topography model.

We will present two independent experiments. First, the quality of the error estimation was analyzed, using an in-depth Monte-Carlo analysis applied to synthetic topography. The implications from these synthetic tests were then verified using two real-world data sets.

6.3.1 Monte-Carlo-Analysis

We tested the error estimation using a Monte-Carlo approach on synthetic random topography. Accurate estimation of error is crucial for the success of the adaptive algorithm. Clearly, it would be best if the error estimate was perfectly accurate. Inaccurate estimation of error could cause two problems. If the error is underestimated, the algorithm will accept a resolution that is too crude, possibly returning an inaccurate gravity correction result. Overestimation, on the other hand, will slow the algorithm down. Since reliability trumps speed, mild overestimation is acceptable, but underestimation should be avoided.

To test how well the error is estimated, it is sufficient to study a small region, say no more than 32×32 cells. This small area is representative for the kind of decision the algorithm has to make during the calculation. By changing the nominal resolution of the test grid, scenarios representative for the higher layers of the quadtree can be tested. Thus, if the error estimation is acceptable for these small areas, it will be acceptable for larger grids as well. Creating synthetic topography for such small number of nodes is straightforward, and by adjusting the parameters controlling how topography is created, a wide range of plausible topographic environments can be studied.

We assume topography can be described by a Gaussian random field, having a constant expectation and an exponential isotropic covariance function:

$$C(d) = Se^{-3d/r}$$

where S is the "sill," r is the "range," and d is the distance between two points. The range r controls the smoothness of the synthetic topography: A higher range means the generated topography is more smooth. The height values of the synthetic topography are h_i. Under these assumptions, h_i have a multivariate normal distribution, from which samples can be drawn using standard techniques.

The parameters used for the Monte-Carlo study are as follows: two cell sizes (grid spacing) were used: 100 m and 10 km. The first resolution roughly corresponds to the resolution of Shuttle Radar Topography Mission (SRTM), which will be used in our real-world data test. The second resolution is typical for resolutions at the starting layer j^*. All grids were square, and the number of cells along each edge was either 4, 8, 16, or 32.

The sill was constant at 0.1 km^2 because we found that it acts as a scaling factor and has little impact on the algorithm's behavior. The range varied between 100 m and 10,000 km. The mean height of topography was 1 km. For each set of parameters (number of cells, size of each cell, range), 1000 random grids were generated. The total grid length is the product of cell size and number of cells per dimension. It is thus equal to the resolution of the grid cells at layer n, i.e., the root of the tree.

Gravity is calculated for a grid of stations placed 1 m above the center of each grid cell. (There is no specific reason for the 1 m station height option—all other station heights would also be feasible.) Note that the number of stations is equal to the number of grid cells. For each station, the gravity effect of layer n, $n-1$, and 0 was computed, giving g_n, g_{n-1}, and g_0. g_0 uses all available topographic information and is hence used as reference value.

g_n is an approximation of g_0, with an error $\epsilon = |g_n - g_0|$. The goal of the adaptive algorithm is to keep ϵ below the user-specified threshold. An estimate of ϵ can be obtained by taking the difference between g_n and g_{n-1}. This gives an estimated error $\tilde{\epsilon} = |g_n - g_{n-1}|$ for g_n. The adaptive algorithm can only function, if $\tilde{\epsilon}$ is not systematically underestimating ϵ.

We define the average success rate as the probability that the estimated error is larger than the true error because we want to avoid underestimation of error. The average success rate is calculated over all 1000 random grids and stations for each set of parameters. It will serve as a metric to evaluate the error estimation.

6.3.1.1 Results

6.3.1.1.1 Influence of the Range on Success Rate

Fig. 6.3 highlights the influence of the geostatistical range of the topographic grid on the average success rate. For low ranges, the average

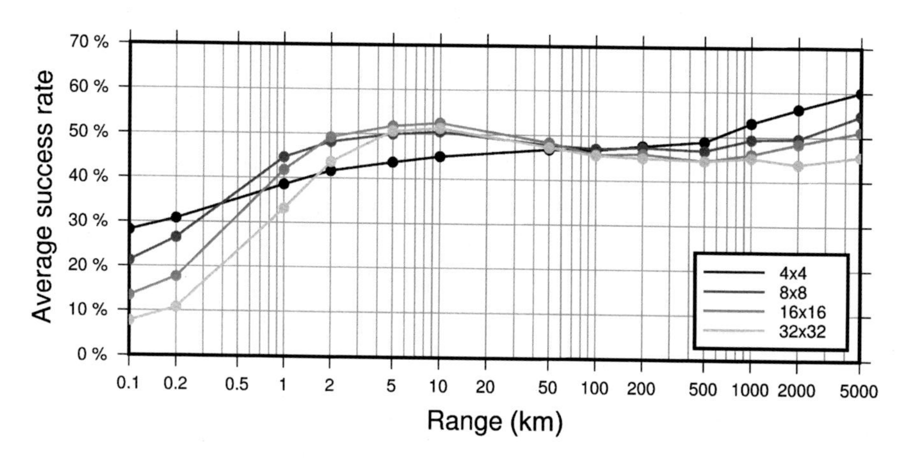

Figure 6.3 Result of the Monte-Carlo test for a cell size of 100 m. As the range of the topographic grid increases, the error estimate becomes more accurate. Once the range of the topography grid is above 10 km, the average success rate stays almost constant at around 50%. For small grids the success rate is higher because the algorithms need to extrapolate over less layers. However, for ranges above 10 km the differences between grid sizes become negligible. Therefore, the range exerts strong influence on the behavior of the adaptive algorithm.

success rate increases strongly with range, reaching a maximum of about 50% for ranges of 10 km. For ranges above 10 km, the success rate decreases again slightly. The success rates are similar for the different grid sizes, except 4×4, if the range is above 5 km.

When the cell size is increased by a factor of 100 to 10 km, the results scale consistently. That is, the maximum success is attained at a range of 1000 km, which is 100 times the threshold observed for the cell size of 100 m. However, the maximum success rate is reduced to 40%.

Results for all cell sizes, grid sizes, and ranges can be combined in an elegant way, by using the ratio of range to total grid length. This ratio will be called χ. Fig. 6.4 shows that χ exerts strong control over the average success rate. For $\chi < 3$, the average success rate increases with χ. This increase tapers off at $\chi = 3$, so that for $\chi > 3$ the success rate changes only slightly with χ. The success rate increases again when the ratio becomes larger than 1000, but such high values of χ are unlikely to appear in real world data.

As a first result, we conclude that the results of the Monte-Carlo method show that the ratio of the range to total grid length has critical impact on the quality of the error estimation. If this ratio is less than three, the adaptive algorithm suffers from systematic underestimation of error. Consequently, the minimum resolution of the adaptive algorithm should be chosen to be no more than one-third of the range of

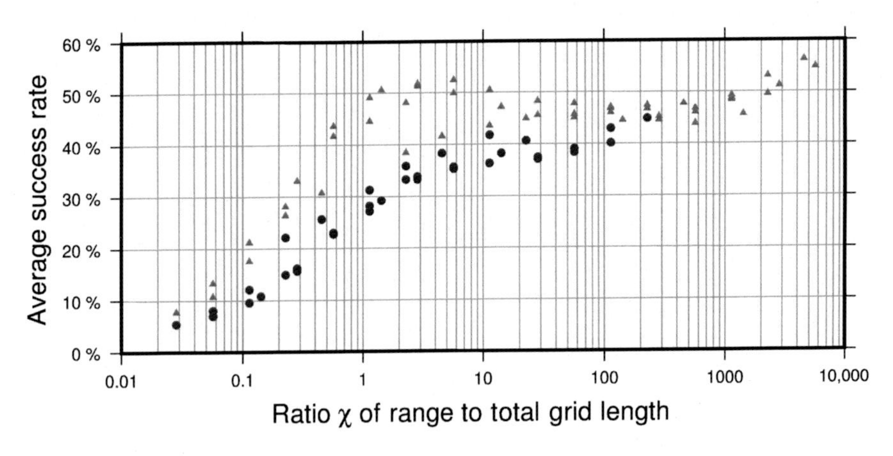

Figure 6.4 Results of the Monte Carlo tests as a function of χ. The blue dots (gray in print versions) belong to a cell size of 10 km, the red triangles to a cell size of 100 m. The behavior of the average success ratio is strongly controlled by χ. For χ larger than three, the average success is almost constant, except for extreme values of χ(>1000).

the topography. Whether or not a success rate of 50% (40%) is enough for practical applications will be tested in the next section.

6.3.2 Real World Data

After the numerical tests by aid of synthetic topography now we chose two real world datasets to test the adaptive algorithm: data which cover

- the Tibetan plateau and
- the globe

6.3.2.1 Tibetan Plateau

The first data set is a rectangular area in the Tibetan plateau (from 30°N to 33.5°N and 85°E to 88.5°E) as test bed for our algorithm (see Fig. 6.5). The area is challenging because the topographic characteristic of landscape varies substantially, showing both vast plains and ragged mountains. Furthermore, there are no oceans in this area, so we can use reprocessed SRTM data with a resolution of 90 m (Jarvis et al., 2008). The grid contains about 16 million data points. Geostatistical analysis of the area indicates constant mean height and an exponential, isotropic covariance function. The sill is around

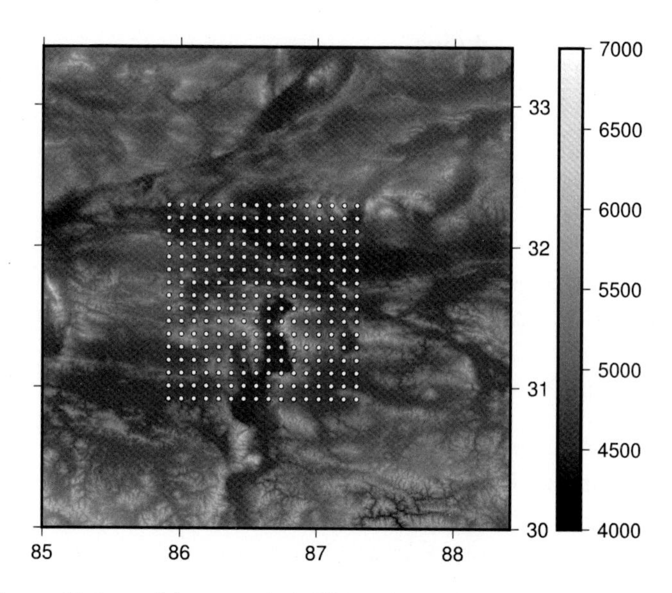

Figure 6.5 Topographical map of the test case in the Tibetan plateau. The color scale gives topographical height in meters. Each blue dot gives the location of a gravity station used in the test. Note that the stations are well removed from the edge of the topographical grid, so edge effects can be ignored (white in print versions).

$0.1\ km^2$, and the range is up to 65 km, which is within the range of values used for the synthetic test. Therefore, the results from the synthetic approach can be applied.

We calculated the gravity effect for a grid of 16×16 stations in the center of the study area. The horizontal distance between stations is about 5 km, so the stations cover an area of roughly 90×90 km. The area covered by station is small compared to the size of the overall topography grid, so edge effects are negligible. The height of the stations is 1 m above topographical height extracted from the SRTM grid (refer also to comment at page 100).

Two experiments were carried out with this data set.

- First, the overall impact of the topography grid's resolution on the accuracy of the obtained gravity value was studied. This provides some guideline for choosing the error tolerance as a function of the grid resolution. However, this neglects other factors such as the measurement accuracy.
- Second, the adaptive algorithm was applied to the Tibetan data set, to how the parameters (error tolerance and maximum cell size) must be chosen to ensure reliability and efficiency.

6.3.2.2 Influence of Resolution

Each layer of the quadtree is a complete topographical grid at a certain resolution. Comparing the gravity effect of a layer with the true gravity value thus gives an estimate of the error incurred by using the resolution of that layer.

Fig. 6.6 shows the root-mean-square error over all 256 stations as a function of the cell size (resolution). For cell sizes below 20 km the error is roughly proportional to $l^{1.06}$, where l is the resolution. If the cell size is above 20 km, the average error changes very little and remains at about 22 mGal[1].

A least-square linear fit for cell sizes below 10 km gives the dashed black line in Fig. 6.6. This relation implies that at a resolution of 90 m, subscale variations cause an error in the order of 0.1 mGal. This represents an absolute lower bound on the errors, because all other causes of error are neglected. Thus, it makes little sense

[1] $1\ mGal = 10^{-5}\ m/s^2$.

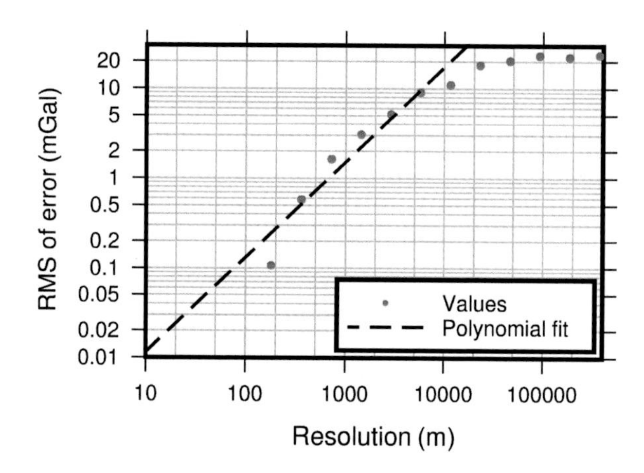

Figure 6.6 Average error relative to a resolution of 90 m. The dashed black line is obtained by fitting a polynomial function to the data points with resolution smaller than 20 km. Note that the RMS error for resolutions of 10 km or more changes very little compared to smaller cell sizes. This qualitative change in behavior is consistent with the results from the synthetic study. Thus, resolution and range of a topographic grid should be strongly connected.

to choose an error tolerance that is much lower than 0.1 mGal for this grid.

Furthermore, Fig. 6.6 corroborates the findings of the synthetic Monte-Carlo study. The error behavior changes qualitatively at a resolution of roughly 20 km, which with a range of ~65 km coincides with $\chi \approx 3$. The explanation is that if the resolution is above 20 km, the gravity effects of the consecutive layers are very similar. Accordingly, if the adaptive calculation begins at one of these layers, the error estimate will be too small systematically, resulting in a reduced success rate.

6.3.2.3 Adaptive Algorithm

We tested the adaptive algorithm using a set of different parameters. The error tolerance varied between 0.01 and 100 mGal. The starting layer j^* was between 7 and 11, corresponding to maximum cell sizes of 12 to 190 km (2^7 to 2^{11} times the original grid resolution of 90 m). We calculated the gravity using the adaptive approach for all 256 stations for all combinations of the two parameters.

We calculated the maximum error over all stations. Each line in Fig. 6.7 shows how the maximum error changes as a function of the error tolerance, for a given maximum cell size. Points outside the gray

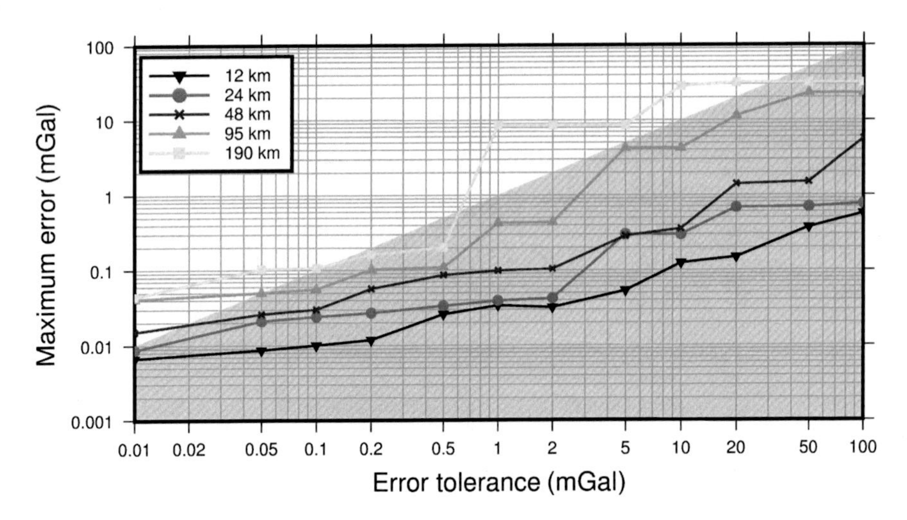

Figure 6.7 Maximum error of all stations as a function of the error tolerance. The shaded gray area is admissible because the maximum error is lower than the error tolerance. Outside of the gray area the accuracy requirement has not been met. Each line is for a different maximum cell size. Only a maximum cell size of 24 km or 12 km is admissible for all error tolerances.

area in Fig. 6.7 exceed the error tolerance. Thus, all lines should ideally lie within the gray area. However, only the curves for 24 and 12 km minimum resolution reliably achieve this. If the error tolerance can be larger than 0.1 mGal, minimum resolutions of 48 or 95 km are also acceptable.

The curve for 190 km shows somewhat erratic behavior: For a tolerance of 10 mGal, it surpasses the error threshold by a factor of three, while for a much stricter tolerance of 0.5 mGal it does not. This behavior can be explained by the fact that the maximum error is determined by a single station. Since we have only used 256 stations, it is mere coincidence that the achieved error drops. Still, the algorithm is not reliable with this maximum resolution.

This observation is consistent with our synthetic tests. As the range of the topographic grid is 65 km, we expect that the behavior of the algorithm changes at around 20 km (corresponding to $\chi = 3$). However, for medium error tolerances, the algorithm's performance is satisfactory even with much more lenient constraints on the maximum cell size (i.e., 95 km).

To judge the efficiency of the adaptive approach, we compare the average number of tesseroids used per station with the total number of grid cells. The number of tesseroids counts tesseroids from all layers of the quad tree, not only those at the highest resolution/lowest layer. This gives an adequate metric for the adaptive approach because it accounts for the overhead of the adaptive algorithm. Also, the number of tesseroids used should be representative for the results obtained using any elementary body.

For a maximum cell size of 190 km, the number of tesseroids depends polynomially on the error tolerance: To decrease the error by a factor of 10, four times as many tesseroids are needed for calculations. As expected, enforcing smaller tesseroids by reducing the maximum cell size, increases the number of required tesseroids (Fig. 6.8).

However, for small error tolerances, the curves for different minimum resolutions converge. The maximum cell size also defines a minimum number of tesseroids. Thus, the number of tesseroids flattens out for large error tolerances (Fig. 6.8). Furthermore, the maximum number of tesseroids used per station was around 570,000. This is equal to 3.5% of the total number of cells in the grid. Even with the strictest accuracy requirements, the algorithm only needs to calculate 3.5% the number of tesseroids compared to the naïve approach.

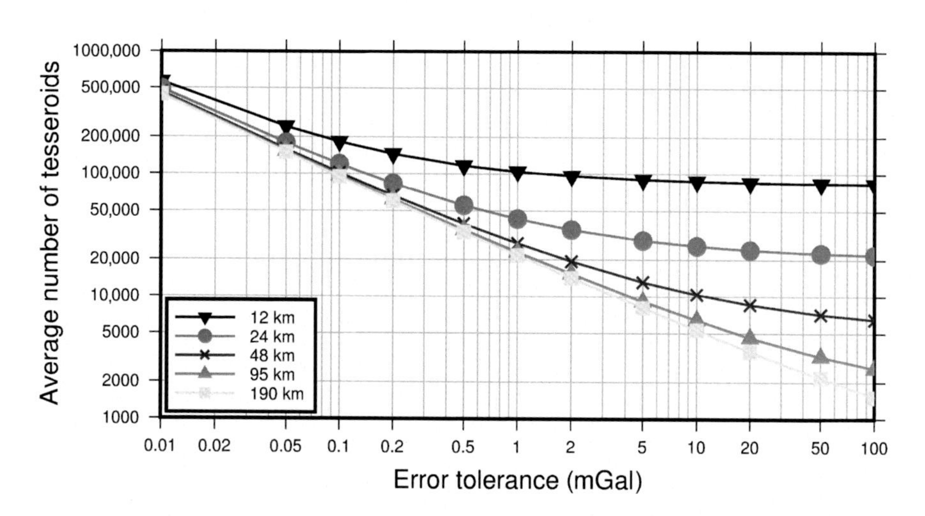

Figure 6.8 Average number of tesseroids used per station. Each line is for a different maximum resolution. Note that the number of tesseroids increases roughly polynomially with the error tolerance. However, the curves for small maximum cell sizes are flatter for larger error tolerances, because they have a higher minimum number of tesseroids. As the error tolerance becomes smaller the curves get closer together (although this effect is exaggerated by the logarithmic scaling). Hence, for strict error tolerances, the chosen maximum cell size has a small effect on the numeric efficiency.

6.3.2.4 Global Calculation

As second data set, we used ETOPO 1 (Amante and Eakins, 2009) a global topographic data set with a resolution of 1 arc min. Thus, there are about 233 million grid points in total. Geostatistical parameters range and sill are about 3000 km and 7 km^2, respectively, for this grid. Accordingly, minimum resolution was set to 1000 km. Gravity was calculated on a global grid of gravity station at a grid spacing of 1°. Station height is 1 m above topography onshore and 1 masl offshore. An error tolerance of 1 mGal was used for the adaptive calculation.

Fig. 6.9 shows the total mass correction obtained using the adaptive algorithm. A strong correlation with topography was expected, because the "Bouguer"-part of mass correction has a dominating influence. Our results agree with the global topographic effect calculated by (Balmino et al., 2012). In Fig. 6.10, the DRE has been isolated. This finding agrees well with the results of Mikuška et al. (2006) and underlines the need of taking distant topography into account. *However note that Mikuska et al. use a different sign convention compared to the results in* Fig. 6.10.

The gain in speed due to the adaptive algorithm is demonstrated by the number of tesseroids used during the calculation (see Fig. 6.11). At most 400,000 tesseroids are used, which is equivalent to 0.17% of the data points. These extreme values are limited to relatively small areas (i.e., the Tibetan plateau, the Eastern Pacific rim), and in most regions considerably less tesseroids are needed.

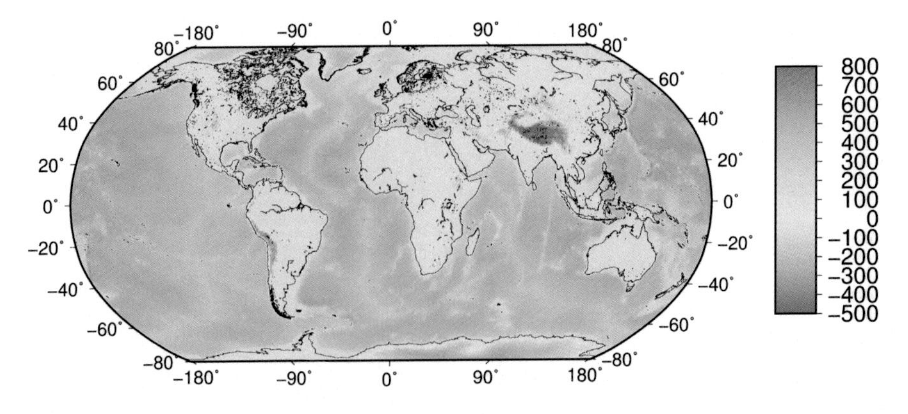

Figure 6.9 Mass correction in mGal 1 m above topography (sea level) calculated using ETOPO 1.

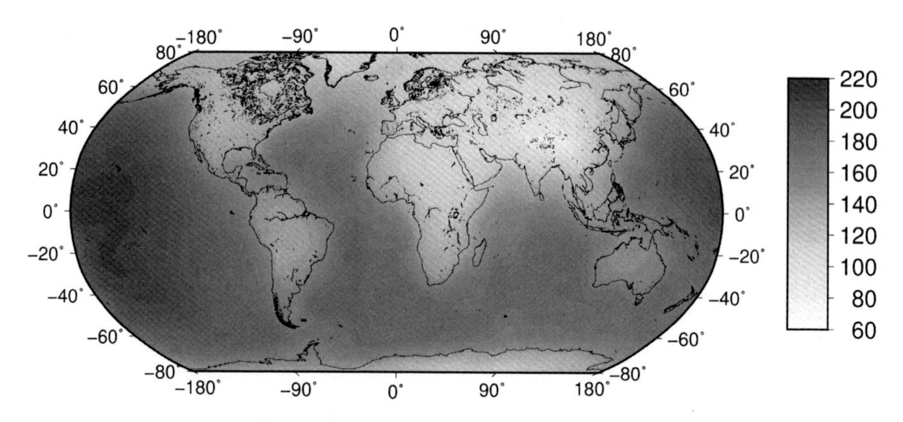

Figure 6.10 Distant Relief Effect in mGal 1 m above topography calculated using the ETOPO 1 model.

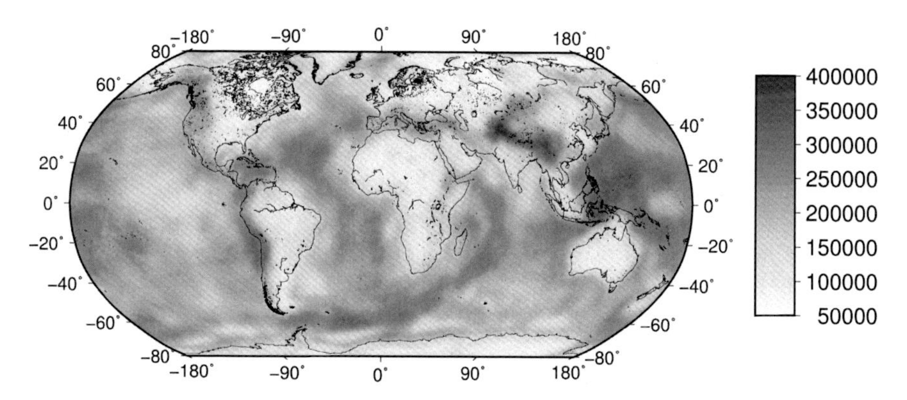

Figure 6.11 Number of tesseroids used, during the adaptive calculation with ETOPO 1. Note that ETOPO 1 model contains 233 million data points in total. Thus, the number of calculated tesseroids is less than 0.17% of the original number of grid cells. In most areas considerably less tesseroids are used. For instance, most of Africa requires less than 50,000 tesseroids, equivalent to 0.02% of original grid cells.

6.4 DISCUSSION

Based on our synthetic and real world test cases, we conclude that the adaptive algorithm is both reliable and efficient.

The "Tibetan" test using high-resolution (90 m) topographic data demonstrates that our adaptive approach is capable of reproducing the nonadaptive result with an accuracy 0.01 mGal, using only 3.5% of the grid cells. This precision is well below the uncertainties associated with modeling assumptions and measurement accuracy at continental scale. If the error tolerance is more lenient, the adaptive algorithm uses less grid cells, so that less than 1.3% of the grid cells are required for an accuracy of 0.1 mGal.

How much the number of tesseroids can be reduced depends on the following factors:

- The number of cells in the original grid, and thus the resolution and horizontal extent of the original grid. The amount of reduction is proportional to the number of cells.
- The minimum resolution sets a lower limit for the number of tesseroids.
- The error tolerance. Decreasing the error by a factor of 10 increases the number of tesseroids by 4.
- The geostatistical properties, most notably the range, affect the required resolution both directly and indirectly through the required minimum resolution.

The observed speed-ups were in the order of 30 (for the area of the Tibetan plateau) to 1000 (for global calculations).

Reliable behavior of the adaptive algorithm was ensured by requiring a maximum cell size. Using synthetic tests, we found that the relation between total grid length and geostatistical range exerts critical influence on the behavior of the algorithm. The algorithm estimates the error made by a coarse resolution effectively only if the ratio of range to total grid length is less than 3. Total grid length corresponds to the maximum cell size in the real world tests. Hence, the maximum cell size can be chosen according to the range of the study area. The validity of this approach was demonstrated in the "Tibetan" test case.

However, studies of synthetic cases are subject to a number of limitations. First, our geostatistical model was rather simple. We neglected anisotropy and spatial variability of mean height, range or sill; and used only a single (exponential) shape of the covariance function. Furthermore, we limited our test to Gaussian random fields, so we did not consider higher order moments, which are often necessary for an adequate description of real topography (Goff and Jordan, 1988). Second, the generated synthetic grids had a maximum dimension of 32×32, which is small compared with the extent of actual topography data. However, since the results for the Tibetan test case are consistent with the synthetic observations, we maintain that the geostatistical range can be used as a basis for choosing the required maximum cell size.

The efficiency of the algorithm is clearly demonstrated by the "Tibetan" test case and the results of the global calculation. As measure of efficiency, we use the number of tesseroids used compared to the original number of cells in the grid. We find that the adaptive algorithm causes some numerical overhead by constructing the quadtree and estimating the error, but it is still considerably faster than the naïve approach.

Adaptive algorithms provide an elegant way of combining high resolution and large lateral extent of topographic data. Thus, both distant and local gravity contributions of the topographic grid can be included. This makes them very attractive for surveys at regional scale. Furthermore, our adaptive approach is flexible. It can thus be directly applied to satellite or airborne gravity data as well as gravity gradient measurements.

ACKNOWLEDGMENTS

This study contributes to research which was done in the Priority Programme SPP 1257 "Mass transport and mass distribution". It has been financed by the Deutsche Forschungsgemeinschaft (DFG), grant (GO 380/27-1/2). We also thank A. Cogbill and one anonymous reviewer whose comments helped substantially to improve a first version of this manuscript.

REFERENCES

Amante, C. and B.W. Eakins (2009). ETOPO1 1 Arc-Minute Global Relief Model: Procedures, Data Sources and Analysis.

Balmino, G., Vales, N., Bonvalot, S., Briais, A., 2012. Spherical harmonic modelling to ultra-high degree of Bouguer and isostatic anomalies. J. Geod. 86 (7), 499–520. Available from: http://dx.doi.org/10.1007/s00190-011-0533-4.

Bouman, J., Ebbing, J., Meekes, S., Fattah, R.A., Fuchs, M., Gradmann, S., et al., 2015. GOCE gravity gradient data for lithospheric modeling. Int. J. Appl. Earth Observ. Geoinform. 35, 16–30.

Bullard, E.C., 1936. Gravity measurements in East Africa. Phil. Trans. R. Soc. A: Math. Phys. Eng. Sci. 235 (757), 445–531. Available from: http://dx.doi.org/10.1098/rsta.1936.0008.

Cogbill, A., 1990. Gravity terrain corrections using digital elevation models. Geophysics 55, 102–106.

Gander, W., Gautschi, W., 2000. Adaptive quadrature – revisited. Bit Numer. Math. 40 (1), 84–101. Available from: http://dx.doi.org/10.1023/A:1022318402393.

Goff, J.A., Jordan, T.H., 1988. Stochastic modeling of seafloor morphology: inversion of sea beam data for second-order statistics. J. Geophys. Res. 93 (B11), 13589. Available from: http://dx.doi.org/10.1029/JB093iB11p13589.

Grombein, T., Seitz, K., Heck, B., 2013. Optimized formulas for the gravitational field of a tesseroid. J. Geod. 87 (7), 645–660. Available from: http://dx.doi.org/10.1007/s00190-013-0636-1.

Hammer, S., 1939. Terrain corrections for gravimeter stations. Geophysics 4 (3), 184–194.

Hayford, J.F., Bowie, W., 1912. The Effect of Topography and Isostatic Compensation Upon the Intensity of Gravity. U.S. Coast and Geodetic Survey, Special Publication No. 10, 132 pp.

Heck, B., Seitz, K., 2007. A comparison of the tesseroid, prism and point-mass approaches for mass reductions in gravity field modelling. J. Geod. 81 (2), 121–136. Available from: http://dx.doi.org/10.1007/s00190-006-0094-0.

Holzrichter, N., 2013. Processing and Interpretation of Satellite and Ground Based Gravity Data at Different Lithospheric Scales, PhD Thesis. Kiel University, Kiel.

Jarvis, A., H.I. Reuter, A. Nelson, and E. Guevara (2008). Hole-Filled SRTM for The Globe Version 4, available from the CGIAR-CSI SRTM 90m Database (http://srtm.csi.cgiar.org)., http://srtm.csi.cgiar.org.

LaFehr, T.R., 1991. Standardization in gravity reduction. Geophysics 56 (8), 1170–1178. Available from: http://dx.doi.org/10.1190/1.1443137.

Mikuška, J., Pašteka, R., Marušiak, I., 2006. Estimation of distant relief effect in gravimetry. Geophysics 71 (6), J59–J69. Available from: http://dx.doi.org/10.1190/1.2338333.

Nabighian, M.N., Ander, M.E., Grauch, V.J.S., Hansen, R.O., LaFehr, T.R., Li, Y., et al., 2005. Historical development of the gravity method in exploration. Geophysics 70 (6), 63ND–89ND. Available from: http://dx.doi.org/10.1190/1.2133785.

Nagy, D., Papp, G., Benedek, J., 2000. The gravitational potential and its derivatives for the prism. Journal of Geodesy 74, 552–560.

Samet, H., 1990. The Design and Analysis of Spatial Data Structures. Addison-Wesley Series in Computer Science, Addison-Wesley, Reading, MA, p. 493.

Szwillus, W., 2014. Ein adaptives Verfahren zur Berechnung der Massenkorrektur, Master Thesis. Kiel University, Kiel.

CHAPTER 7

National Gravimetric Database of the Slovak Republic

Pavol Zahorec[1], Roman Pašteka[2], Ján Mikuška[3], Viktória Szalaiová[4], Juraj Papčo[5], David Kušnirák[2], Jaroslava Pánisová[6], Martin Krajňák[2], Peter Vajda[6], Miroslav Bielik[2] and Ivan Marušiak[3]

[1]Slovak Academy of Sciences, Banská Bystrica, Slovak Republic [2]Comenius University, Bratislava, Slovak Republic [3]G-trend s.r.o., Bratislava, Slovak Republic [4]Geocomplex, a.s., Bratislava, Slovak Republic [5]Slovak University of Technology, Bratislava, Slovak Republic [6]Slovak Academy of Sciences, Bratislava, Slovak Republic

7.1 INTRODUCTION

The territory of the Slovak Republic (except the inaccessible area of the Tatra Mountains) is covered by regional gravity measurements in the scale 1:25,000, which represents 3–6 points/km^2. The measurements were realized during a long period from the 1950s up to the 1990s (Fig. 7.1). The project goal was to create a high definition gravity map for mineral exploration and basic geologic interpretations. Various types of gravity meters were used during the data acquisition time period (GAK PT, Worden, Canadien CG-2, Scintrex CG-3M). Different approaches to complete Bouguer anomaly (CBA) calculation were used, including different normal field formulas, different equations for "Bouguer" correction and atmospheric correction, as well as various methods of the terrain correction estimation. A complete recalculation of the entire database was performed in the frame of the earlier project *Atlas of geophysical map and lines* (Grand et al., 2001). Several hundreds of random error points (with errors in their heights or positions) were identified—these points have been removed from the final Bouguer anomaly evaluation. Systematic errors in the outer zone terrain corrections (outer zone T3 from 5.24 to 166.7 km) were eliminated (Fig. 7.2). However, large errors in the inner zone terrain corrections were still expected because of the inaccurate elevation models available that time (2001).

In addition to the regional gravity measurements, a quantity of local detailed gravity surveys was realized in Slovakia during the last

Understanding the Bouguer Anomaly. DOI: http://dx.doi.org/10.1016/B978-0-12-812913-5.00006-3

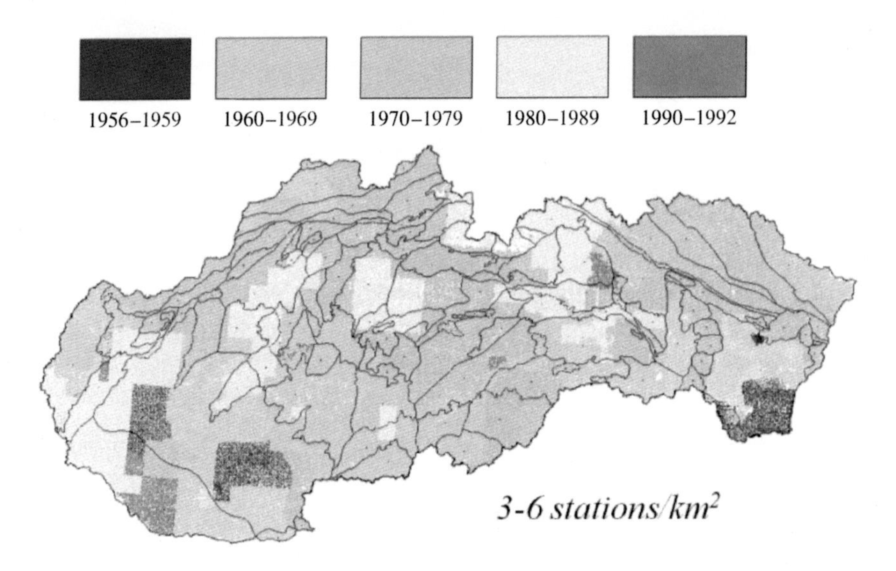

Figure 7.1 Time period of regional gravity database measurements.

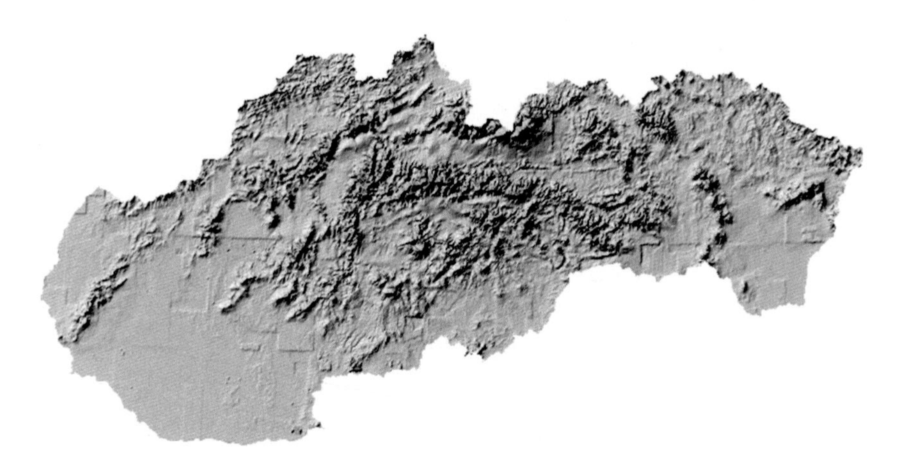

Figure 7.2 Shaded relief map of the older version (1992) of the outer zone terrain corrections T3 from 5.24 to 166.7 km (range from −0.67 to 6.46 mGal for the density 1.0 g/cm^3); after Grand et al. (2004). The pattern following the map sheets margins is clearly visible.

40 years. These data were not incorporated into the unified database before the current project.

7.2 COMPILATION OF INTEGRATED GRAVITY DATABASE

In the frame of actual research project APVV-0194-10 *Bouguer anomalies of new generation and the gravimetrical model of Western*

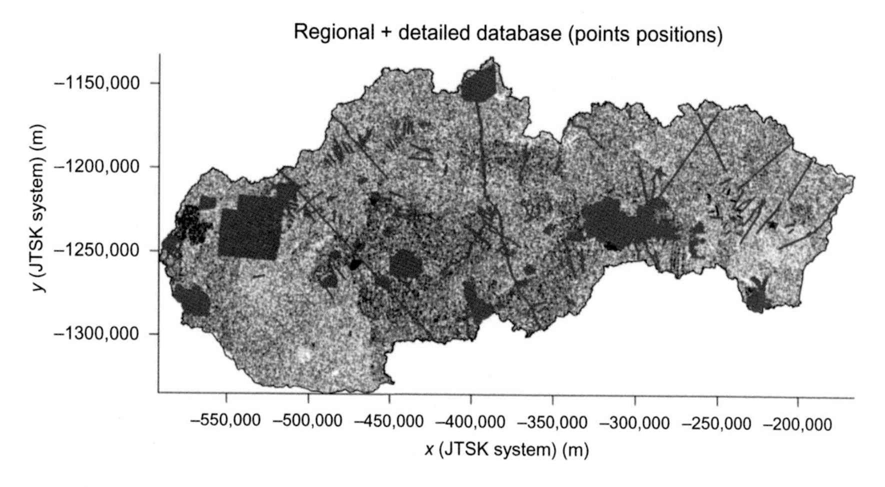

*Figure 7.3 Location of regional (*black*) and detailed (*blue *(gray in print versions)) gravity points.*

Carpathians, all available gravity data in Slovakia were integrated into the unified gravimetric database. The existing regional gravity database (212,478 points) was supplemented with 107,437 detailed gravity points, as seen in Fig. 7.3.

The detailed gravity database is composed of approximately 100 various local projects for the oil-and-gas, mineral and geothermal exploration, basic geology and environmental applications (mainly from archives of Geocomplex, Ltd. and other cooperating organizations in the frame of the actual project). These measurements were realized during the last 40 years, from the 1970s until today. Our analysis indicates that especially the older datasets (measured before 1990s) contain random or systematic errors in the positions, heights, and/or gravity values. Several hundreds of evidently incorrect points (scattered points, local points with the extreme height differences compared with the elevation model, etc.) were excluded from this detailed database. We have tested the accuracy of the original coordinates, which were derived manually from maps in the scale 1:25,000, because they were not surveyed in the field during the gravity measurements. We have digitized about 50 original map sheets and compared the acquired point coordinates at 8797 points with the original ones. As can be seen in Fig. 7.4, there are some points with position differences of hundreds of meters (which can lead to large errors in Bouguer anomaly).

However, on average, the position differences are on an acceptable level. We note that this test does not deal with the real

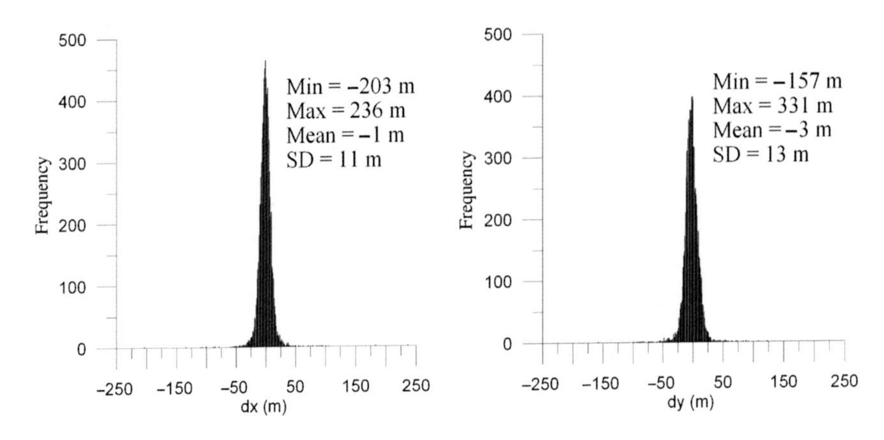

Figure 7.4 Differences between digitized and original manually derived coordinates at 8797 points in x-*coordinate (left) and* y-*coordinate (right).*

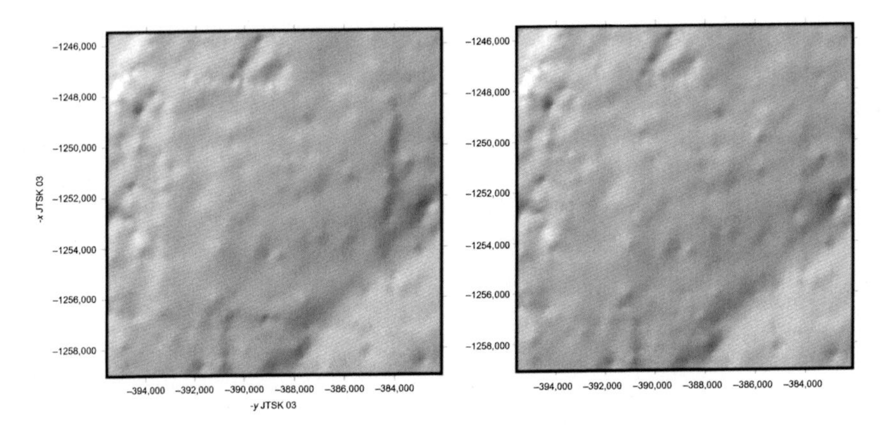

Figure 7.5 Shaded relief map of Bouguer anomalies with well visible systematic error of about +0.6 mGal within a single map sheet before (left) and after (right) correction.

accuracy of the gravity-point position, but with the accuracy of their manually derived coordinates.

Several hundreds of points with systematic errors in the measured gravity were corrected on a basis of the control field measurements as a part of this data merge project. An example of such systematic error within a single map sheet is shown in Fig. 7.5. As we found during the verification measurements, the measured gravity on points within this map sheet (probably by the same operator) was systematically incorrect to about 0.6 mGal.

In addition to performing visual qualitative control and identifying so-called bull-eye anomalies, all data were also analyzed by the quantitative criteria. We compared the heights of measured points with the actual detailed digital elevation model (DEM) of Slovakia DMR-3 (Topographic Institute, 2012). Points with the height residuals larger than ± 40 m were considered to be probably erroneous points (this limit was estimated on the basis of DEM quality control; see the next section). We also compared the calculated CBA values with the recalculated regional CBA map of Slovakia (the same processing was used). Points with residuals larger than ± 5 mGal were also considered to be probably erroneous points. We left such points in the database (marked by a special quality code), so that they can be further scrutinized during subsequent interpretation.

Detailed gravity data were archived in various coordinate systems as well as gravimetric reference systems, and therefore they had to be unified. The transformation among used coordinate systems (S-JTSK (JTSK03), S-1942, S-1952, ETRS89) was performed using the software Univcol (Marušiak, 2012). S-JTSK (realization JTSK03) is the national coordinate system in Slovakia. It is based on Bessel ellipsoid with the Krovak conical projection in common position. Transformation to the European Terrestrial Reference System 1989 (European version of the global coordinate system known as the International Terrestrial Reference System) is based on the Bursa-Wolf 6-parameter transformation. Transformation parameters are in full compliance with the official instruction of the National Mapping Agency. The official coordinates S-JTSK03 and ETRS89 are archived in the final database. We have analyzed in detail the transformation between the former gravity reference system (Gravity system 1964) and actual Gravity system 1995 (the system is realized by 10 absolute gravity points and 278 points of the basic gravity network), using 28 identical reference points. Differences between these systems vary from − 13.97 to − 13.77 mGal within the territory of Slovakia (Fig. 7.6). We recognized that there is an indication of the dependency between these differences and the gravity value itself. However, for practical reasons we used the average constant value of − 13.84 mGal for the transformation of gravity data from the system 1964 to the system 1995. For comparison, the previous (2001) recalculation and compilation chose a nearly identical value of − 13.80 mGal.

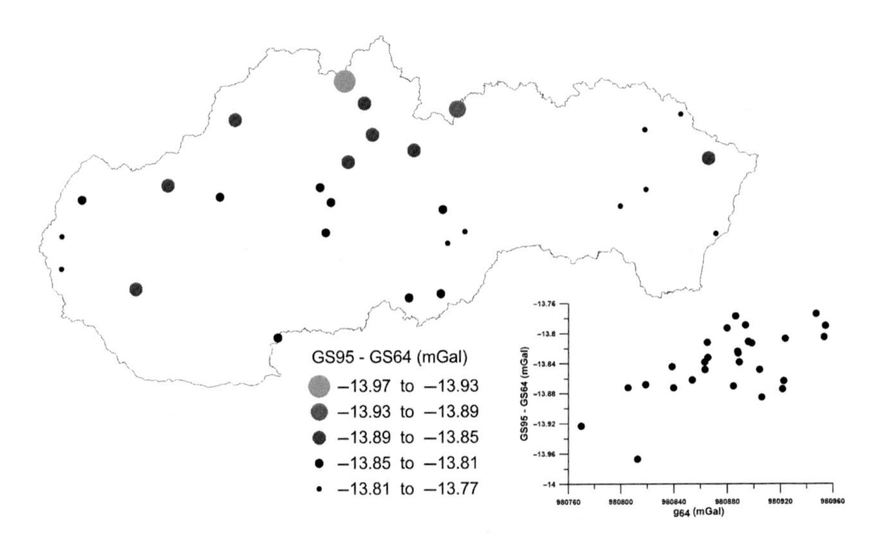

Figure 7.6 Differences between the former and actual gravity reference system on 28 points of basic gravity network. Their dependency on gravity value itself is shown in bottom right.

7.3 NEW GENERATION BOUGUER ANOMALIES

Once the above-mentioned quality control, transformation, and unification aspects were complete, the full database (almost 320,000 points) was reprocessed to the CBA values. One of the most important steps of this process is the precise evaluation of the terrain corrections, especially in mountainous countries like Slovakia. We used a new program Toposk see Chapter 6, Numerical Calculation of Terrain Correction Within the Bouguer Anomaly Evaluation (Program Toposk) in this book and a well-established approach to divide the calculated area into four zones (Grand et al., 2001): inner zone T1 (0–250 m), intermediate zone T2 (250–5240 m), outer zones T31 (5.24–28.8 km), and T32 (28.8–166.7 km). Different numerical approaches and different digital elevation models with the increasing resolution toward the calculation point are used within particular zones. The concept of interpolated heights of the calculation points within the inner zone T1 was used (instead of the measured ones), which we found as a reasonable approach (Grand et al., 2001; Zahorec, 2015). The new detailed digital elevation model was compiled by combination of DMR-3 and DMR-4 (Topographic Institute, 2012) with shuttle radar topography mission (SRTM) data (Jarvis et al., 2008) outside the Slovak territory. Detailed models DMR-3 as well as DMR-4 are connected to the same Slovak local

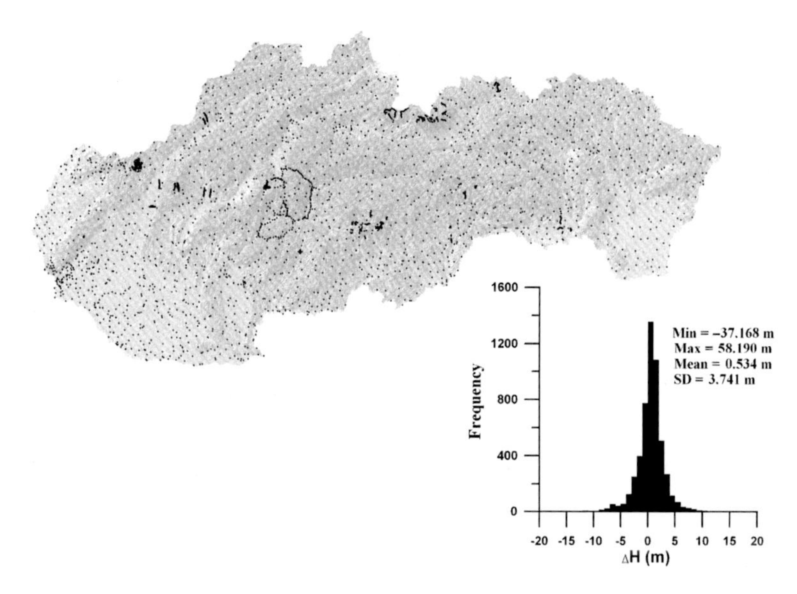

Figure 7.7 Accuracy of the new compiled detail DEM tested on a set of 5299 geodetic and gravity points.

height system—Kronstad-based system Baltic after adjustment (Bpv). Transformation of directly measured ellipsoidal height from global navigation satellite system (GNSS) in ETRS89 (ellipsoid GRS80) to this local height system Bpv is made by the Slovak local quasigeoid DVRM (Klobušiak et al., 2005) related to the same ellipsoid GRS80. Local quasigeoid DVRM is in very good agreement with EGM96 due to its remove-restore creation technique. Differences between them in the area of Slovakia are within the range of -1.5 to 0.3 m, which is definitely better than the precision of the SRTM model.

We tested the new DEM on a set of the state geodetic network points as well as a set of our recent gravity points (measured using GNSS methods). Statistics in Fig. 7.7 show the maximum height differences of several tens of meters even in the highest mountains, which is considerably better than former models, although there are still local errors. The quality of the new compiled DEM has the greatest impact on improving the terrain corrections.

The impact of the new version of terrain corrections is clearly visible at the local CBA map from the Tatra Mountains area (Fig. 7.8). The presence of the local "bull eye" anomalies is visible in the case of the old version of terrain corrections. Maximum differences between the old and new terrain corrections (and consequently the CBA values) are approximately ± 6 mGal for the correction density 2.67 g/cm^3.

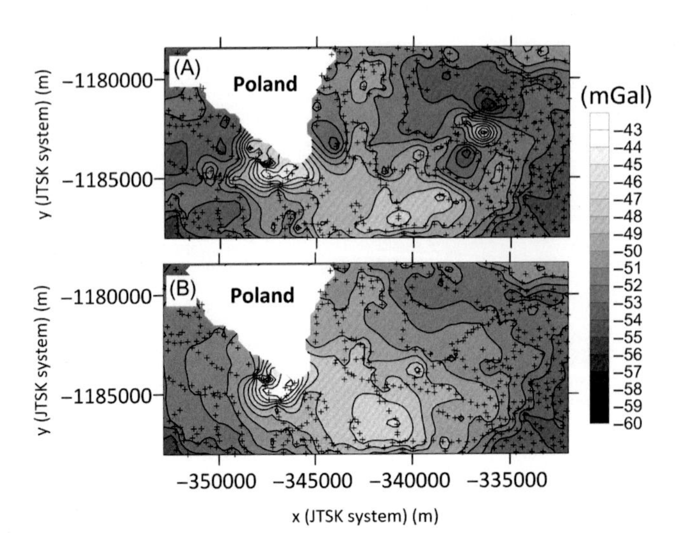

Figure 7.8 Comparison of the local CBA map (a part of the High Tatra Mountains) constructed with the old (A) and new (B) versions of the terrain corrections.

The CBA was calculated following the equation:

$$\text{CBA} = g(P) - \gamma(P_0) - \delta g_F(P) - \delta g_{\text{sph}}(P) + \text{TC}(P) + \delta g_{\text{atm}}(P) \text{ (mGal)}, \tag{7.1}$$

where $g(P)$ is the drift-corrected measured gravity acceleration related to Gravity system 1995, $\gamma(P_0)$ is the normal gravity field (Pizetti-Somigliana formula with GRS80 reference system parameters) on the ellipsoid, $\delta g_F(P)$ is the free air correction term in a second degree approximation (Wenzel, 1985), $\delta g_{\text{sph}}(P)$ is the gravitational effect of truncated spherical layer (Mikuška et al., 2006) with the truncation angle of $1°29'58''$ (corresponding to 166 730 m) and the density 2.67 g/cm^3; this term is known as a Bouguer correction (in spherical approximation), TC(P) is the terrain correction calculated to 166 730 m with the Toposk program (2.67 g/cm^3), and $\delta g_{\text{atm}}(P)$ is the atmospheric correction calculated by the effect of the true atmosphere (Mikuška et al., 2008), using the real topography model and the effect of spherical shell with radially dependent density (Karcol, 2011). The final CBA map is shown in Fig. 7.9.

In addition to the mentioned standard steps of the CBA calculation, we also calculated the distant topography and bathymetry effects (including the ice effect) for the entire database. However, these effects are not incorporated into the anomaly computation for the present analysis; they will be included into the next update of this database. We also recognized that the so-called geophysical indirect effect has a long-wave

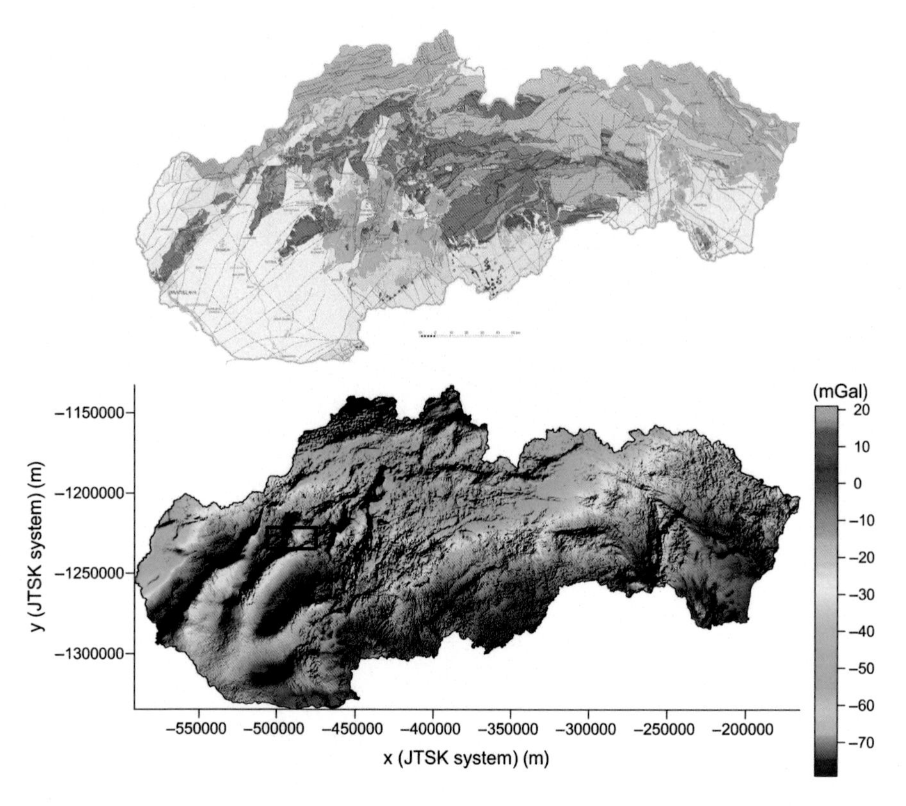

Figure 7.9 Complete Bouguer anomaly map of Slovakia (2.67 g/cm^3) and its comparison with geological map (after Biely et al., 1996). Black rectangle indicates the region with field measurement verification profiles shown in Fig. 7.10.

character which does not affect the local gravity field interpretations. Therefore we use the normal heights, not the ellipsoidal ones.

The final gravimetric database contains the following data (in the American Standard Code for Information Interchange (ASCII) format): point number, year of acquisition, quality code, *x*-coordinate (S-JTSK03), *y*-coordinate (S-JTSK03), longitude (ETRS89), latitude (ETRS89), elevation (Bpv), observed gravity (GS95), near topographic effects NTE1, NTE2, NTE31, and NTE32 (for density 1.0 g/cm^3), terrain corrections T1, T2, T31, and T32 (for density 1.0 g/cm^3), distant topographic effect, distant bathymetric effect, atmospheric effect, CBA (2.67 g/cm^3).

7.4 NEW LINEAR FEATURES RECOGNIZED IN THE BOUGUER ANOMALY MAP

Several new regional linear features were recognized during the CBA map analysis. We have used the detailed field measurements to confirm

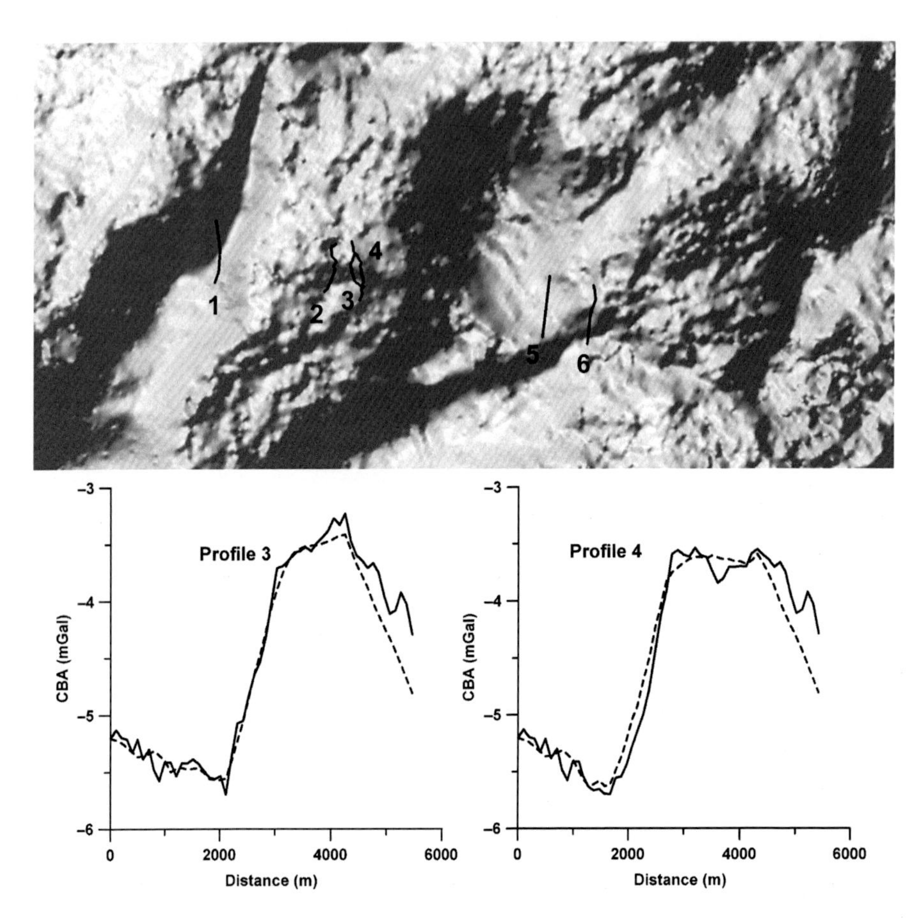

Figure 7.10 In-situ measured control profiles. Comparison between measured values (solid lines) and values interpolated from regional CBA map (dashed lines) along two profiles.

one of them, which is running in the approximate W–E direction (Figs. 7.9 and 7.10). Several profiles were realized in the areas, where the linear structure is visible and not visible, respectively. Comparison of the interpolated CBA values with the measured CBA values along two verification profiles (from area where the structure is clearly visible in the map) is shown in Fig. 7.10. As can be seen, the in-situ measurements confirm the sharp gradient in the CBA values along the properly situated profiles intersecting the linear regional structure. This linear feature has an unexpected strike, and its geological interpretation is still under discussion.

7.5 NEW SOFTWARE FOR THE RECONSTRUCTION OF THE GRAVITY FIELD FROM THE BOUGUER ANOMALY MAP

A new software solution CBA2G_SK (Marušiak et al., 2015) for the recomputation of the gravity acceleration from the newly completed Bouguer anomaly grid was developed. The goal is to estimate the gravity acceleration everywhere within the Slovak territory with the highest possible accuracy. The recalculation process is defined by the reverse equation in regard of the CBA:

$$g(P) = \mathrm{CBA} + \gamma(P_0) + \delta g_F(P) + \delta g_{\mathrm{sph}}(P) - \mathrm{TC}(P) - \delta g_{\mathrm{atm}}(P)\ (\mathrm{mGal}), \quad (7.2)$$

(explanations of symbols are given below Eq. 7.1).

The same DEMs and the zone division to four zones are used as within the original CBA calculation (user interface of the new software can be seen in Fig. 7.11). Main utilization of this software is for more

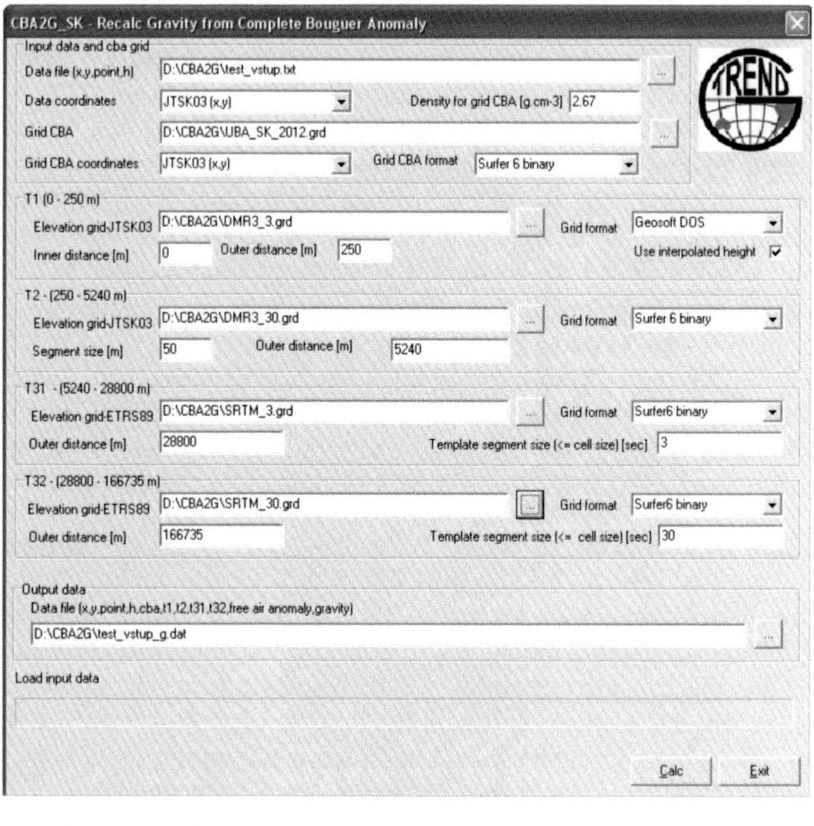

CBA2G_SK - Recalc Gravity from Complete Bouguer Anomaly
Input data and cba grid
Data file (x,y,point,h) D:\CBA2G\test_vstup.txt
Data coordinates JTSK03 (x,y) Density for grid CBA [g.cm-3] 2.67
Grid CBA D:\CBA2G\UBA_SK_2012.grd
Grid CBA coordinates JTSK03 (x,y) Grid CBA format Surfer 6 binary
TREND
T1 (0 - 250 m)
Elevation grid-JTSK03 D:\CBA2G\DMR3_3.grd Grid format Geosoft DOS
Inner distance [m] 0 Outer distance [m] 250 Use interpolated height
T2 - (250 - 5240 m)
Elevation grid-JTSK03 D:\CBA2G\DMR3_30.grd Grid format Surfer 6 binary
Segment size [m] 50 Outer distance [m] 5240
T31 - (5240 - 28800 m)
Elevation grid-ETRS89 D:\CBA2G\SRTM_3.grd Grid format Surfer6 binary
Outer distance [m] 28800 Template segment size (<= cell size) [sec] 3
T32 - (28800 - 166735 m)
Elevation grid-ETRS89 D:\CBA2G\SRTM_30.grd Grid format Surfer6 binary
Outer distance [m] 166735 Template segment size (<= cell size) [sec] 30
Output data
Data file (x,y,point,h,cba,t1,t2,t31,t32,free air anomaly,gravity)
D:\CBA2G\test_vstup_g.dat
Load input data
Calc Exit

Figure 7.11 User interface of the new software CBA2G_SK.

accurate determination of the normal heights within the leveling networks when the in-situ measured gravity values are not available. The tests on the State Gravimetric Network points (these points were not included in the new CBA map calculation; they serve as control points) show maximum differences of about ± 1 mGal between the measured and calculated gravity values. These results confirm the accuracy of the new CBA map of the Slovakia as well as the correctness of the new software algorithm.

7.6 CONCLUSIONS

Large volume gravity dataset covering Slovakia has been reprocessed during the research project described herein. The former regional gravimetric database (212,478 points) was supplemented with the detailed gravity data (107,437 points) covering the entire territory of Slovakia. The new integrated database was qualitatively and quantitatively analyzed, and many random and systematic errors in coordinates, heights, and gravity values were detected and corrected. Detailed analysis of the transformation between previous and actual reference gravity systems was performed, and the constant value of -13.84 mGal (instead of the previously used -13.80 mGal) was used for the transformation between previous and new systems. Close attention was paid to the terrain correction calculation during the "CBA" calculation process. Newly developed software Toposk (see Chapter 5 in this book) was used for terrain corrections, which included the concept of interpolated heights of the calculation points within the innermost zone T1 (0–250 m). New detailed digital elevation model was compiled by combination of local detailed models and SRTM data. Results of testing demonstrated the high quality of this DEM even in the highest mountains for the regional surveys. Differences between the previous and new versions of the terrain corrections reached ± 6 mGal for the correction density of 2.67 g/cm^3.

New regional linear structures were identified on the CBA map and consequently verified by the detailed gravity measurements. A new software solution for estimation of the gravity acceleration from the CBA values was also developed. We believe that the new CBA map will provide a higher reliability for the geological interpretations, as well as for the geodetic applications. Process of improvement of the current gravimetric database will continue in the future as new measurements and more accurate DEM models become available.

ACKNOWLEDGMENTS

The authors are grateful to the Slovak Research and Development Agency APVV (grants No. APVV-0194-10, APVV-0827-12 and APVV-0724-11) and the Slovak Grant Agency VEGA (grants No. 1/0141/15 and 2/0042/15) for support of the presented work. We thank Serguei Goussev and John Bain for reviewing the manuscript.

REFERENCES

Biely A., Bezák V., Elečko M., Kaličiak M., Konečný V., Lexa J., et al., 1996: Geological Map of Slovakia, Simplified Version, Scale 1:500 000. State Geological Institute Dionýza Štúra Bratislava, Slovakia.

Grand, T., Šefara, J., Pašteka, R., Bielik, M., Daniel, S., 2001. Atlas of Geophysical Maps and Profiles. Part D1: Gravimetry. Final Report. State Geological Institute, Bratislava, MS Geofond (in Slovak).

Grand, T., Pašteka, R., Šefara, J., 2004. New version of terrain correction in the Slovak regional gravity database. Contrib. Geophys. Geod. 34, 315–337.

Jarvis, A., Reuter, H.I., Nelson, A., Guevara, E., 2008. Hole-filled SRTM for the Globe Version 4, Available from the CGIAR-CSI SRTM 90m Database: http://srtm.csi.cgiar.org.

Karcol, R., 2011. Gravitational attraction and potential of spherical shell with radially dependent density. Stud. Geophys. Geod. 55/1, 21–34.

Klobušiak, M., Leitmanová, K., Ferianc, D., 2005. Realization of obligatory transformation between national coordinates and height reference system into ETRS89. Proceedings of the International Conference Tatry 2005 (in Slovak).

Marušiak, I., 2012. Univcol 3.5 (Universal Columns Calculator). G-trend Ltd., Bratislava.

Marušiak, I., Mikuška, J., Papčo, J., Zahorec, P., Pašteka, R., 2015. CBA2G_SK (Complete Bouguer Anomaly To Gravity), Program for Calculation of the Gravity Acceleration from Complete Bouguer Anomaly, Program Guide. Manuscript, G-trend Ltd, in Slovak.

Mikuška, J., Pašteka, R., Marušiak, I., 2006. Estimation of distant relief effect in gravimetry. Geophysics 71, J59–J69.

Mikuška, J., Marušiak, I., Pašteka, R., Karcol, R., Beňo, J., 2008. The effect of topography in calculating the atmospheric correction in gravimetry. SEG Las Vegas 2008 Annual Meeting, expanded abstract, 784–788.

Topographic Institute, 2012. Digital Terrain Model Version 3 (Online). http://www.topu.mil.sk/14971/digitalny-model-reliefu-urovne-3-%28dmr-3%29.php.

Wenzel, H.G., 1985. Hochauflösende Kugelfunktionsmodelle für das Gravitationspotential der Erde. Wiss. Arb. Fachr. Vermessungswesen Univ, Hannover, Nr. 137.

Zahorec, P., 2015. Inner zone terrain correction calculation using interpolated heights. Contrib. Geophys. Geod. 45/3, 219–235.

CONCLUSIONS

Roman Pašteka[1], Ján Mikuška[2] and Bruno Meurers[3]
[1]Comenius University, Bratislava, Slovak Republic [2]G-trend, s.r.o., Bratislava, Slovak Republic
[3]University of Vienna, Vienna, Austria

Developments during the last decade show that large discoveries from exploration and production of conventional nonrenewable energy sources are becoming more elusive and expensive. This should encourage the increased utilization of nonseismic methods (mainly the potential field methods—gravimetry, magnetometry, and electromagnetic methods) aiming at reducing average exploration costs. The search for higher effectiveness is valid not only for energy and mineral resources exploration, but also in the broader study of the Earth's crust and lithosphere structure. From this perspective, there exist several important challenges in the area of potential fields: acquisition, processing, and interpretation—some of them are well analyzed in Li and Krahenbuhl (2015). In general, we state that when performing accurate gravity field interpretation and modeling, we must introduce reliable corrections to the acquired data in order to exclude all nongeological signals.

The chapters presented in this book are intended to contribute to the scientific developments connected with terrestrial gravity data processing, mainly with the discussion of the Bouguer gravity anomaly definition. Although the evaluation of Bouguer anomaly values is one of the most important (and standard) steps in anomalous gravity field representation within applied geophysics, there still exist open questions and conflicting aspects. Of course, the presented themes do not cover all the actual problems, connected with precise and correct Bouguer anomaly evaluation (many of those themes are well described in several textbooks, published during the last years, e.g., LaFehr and Nabighian, 2012; Hinze et al., 2013; Fairhead, 2016). Instead, our work in these chapters focuses on several specific aspects, as deemed important by the editors of this volume.

Understanding the Bouguer Anomaly. DOI: http://dx.doi.org/10.1016/B978-0-12-812913-5.00016-6

The most important outputs of this publication are as follows:

- At global scales, the Bouguer gravity anomaly evaluation should be based on ellipsoidal heights. Only an approach with this vertical datum yields the correct evaluation of the gravitational effect of all masses below the topography surface that differ from the density of the reference Earth (Chapter 2, The Physical Meaning of Bouguer Anomalies—General Aspects Revisited). This was pointed out also by the authors of one of the important papers for the Society of Exploration Geophysicists (SEG) community (Hinze et al., 2005). Comment: For some regional and the majority of local surveys, a sea level datum, which is often used conventionally, can be still used as the vertical datum, without undue errors.
- At local and regional scales, the Bouguer gravity anomaly can be regarded to be harmonic in planar approximation everywhere above and on the topo-surface (Chapter 2, The Physical Meaning of Bouguer Anomalies—General Aspects Revisited). The scalar representation of the anomalous gravitational acceleration in Bouguer anomalies is justified for local and regional studies, but it can be associated with considerable errors at larger scales.
- The truncation of topographical masses during gravity terrain corrections (to the standard distance of 166.7 km) is justified only for local- to regional-scale investigations. However, the distant relief effect is important at large scales and even sometimes at regional scales at specific locations (Chapter 2, The Physical Meaning of Bouguer Anomalies—General Aspects Revisited), which was shown also by Mikuška et al. (2006). To some extent, topographical and ocean masses are compensated by crustal thickening and thinning, respectively. Hence, isostatic far-field effects may have to be considered as well (Chapter 2, The Physical Meaning of Bouguer Anomalies—General Aspects Revisited) as they reduce the distant relief effect effectively as shown by Szwillus et al. (2016).
- From the analysis of historical sources, it follows that Pierre Bouguer himself in fact did not introduce the "Bouguer slab" or "Bouguer plate" correction as it could be generally interpreted (Chapter 3, Some Remarks on the Early History of the Bouguer Anomaly). Similar conclusion holds also for the so-called "free-air" or "Faye" correction—it was not suggested in any of the fundamental works of Faye. The term "Bouguer reduction" is likely correctly attributed to Helmert (1884) who used it for describing his relocation of the measured gravity values to sea level datum—a step which

Bouguer, in fact, neither did nor proposed. However, the authors of Chapter 3, Some Remarks on the Early History of the Bouguer Anomaly, concluded that Pierre Bouguer had indeed laid the foundations of the present-day Bouguer anomaly evaluation.

- In this publication, one important methodical approach is presented (Chapter 4, Normal Earth Gravity Field Versus Gravity Effect of Layered Ellipsoidal Model), namely calculating the normal (theoretical) gravity entering into the Bouguer anomaly evaluation: not as the standard solution of the Laplace equation (e.g., Helmert's or Somigliana's formulae), but calculating it as the gravity effect of a layered ellipsoidal model. Despite the fact that some important unanswered questions remain, this approach could be an alternative way to understand and work with the normal gravity field in the future. The main goal of the authors in Chapter 4, Normal Earth Gravity Field Versus Gravity Effect of Layered Ellipsoidal Model, however, was to initiate a broader discussion, rather than to give any definitive solution.
- In the calculation of the topographic effects (or the terrain corrections) as a part of Bouguer anomaly evaluation, there exist a large number of approaches, using different approximations and divisions of the calculation point vicinity into zones (where the nearest relief plays the most important role). In the approach in Chapter 5, Numerical Calculation of Terrain Correction Within the Bouguer Anomaly Evaluation, four circular zones with outer radii of 250, 5240, 28,800, and 166,730 m, respectively, are used within a newly developed software Toposk. The involvement of a 3D polyhedral body in the nearest calculation zone is the best state-of-the-art approximation. Such a concept of polyhedral bodies or a sum of triangular prisms approximation is supported also by other authors (e.g., Tsoulis, 2003; Cella, 2015). Another important aspect of the presented approach is the concept of interpolated heights of calculation points (instead of measured ones) within the nearest zone.
- Another aspect in the evaluation algorithms for topographic effects is their computational speed. This is linked with modern highly detailed and large-scale digital elevation models (e.g., those originated from laser scanning procedures and LiDAR technologies)—using them can make the terrain correction process time-consuming, although of much higher accuracy. This drawback can be overcome by means of a dynamic refinement of the elevation grids during calculation according to the geostatistical properties of topography in the studied area (Chapter 6, Efficient Mass Correction Using an Adaptive Method).

- As a case study of Bouguer anomalies in the evaluation for larger datasets, a practical example from the Slovak Republic is given in Chapter 7, National Gravimetric Database of the Slovak Republic. The former regional gravimetric database (212,478 points) was supplemented with detailed gravity data (107,437 points), and today, it represents one of the most densely covered Bouguer anomalies map of an individual country in the world. In this chapter, several practical aspects are discussed and one newly detected regional linear structure was verified by detailed in-situ measurements. In addition to this, a new software CBA2G_SK is developed, where a reverse calculation method is realized—from the Bouguer anomaly values the value of gravity acceleration in a calculation point is reconstructed. Such estimated values can play a role in various geophysical and geodetic applications, as discussed in Chapter 7, National Gravimetric Database of the Slovak Republic.

Naturally, in addition to the above-mentioned outcomes, the reader can find other interesting comments and ideas within the seven chapters of our monograph. On the other hand, there are still many open problems and questions, which have not been fully covered by the chapters in this monograph and remain for future discussions:

- Why is the standard radius for topographic effect (or gravity terrain corrections) evaluation equal to approx. 166.7 km (opening half-angle of $1°29'58''$)? We know empirically that this distance is very well selected, but do we really know the physical reasons for it?
- Should the standard formulae for the normal (theoretical) gravity (e.g., Helmert's or Somigliana's formulae) be used, or should we develop a concept of normal gravity using the gravity effect of a well-defined reference layered ellipsoidal model? The discussion was initiated in our monograph in Chapter 4, Normal Earth Gravity Field Versus Gravity Effect of Layered Ellipsoidal Model.
- Regarding the topographical effect evaluation: how we can best account for discrepancies between the measured heights of the calculation points and those coming from their projections on the used digital elevation model? Some solutions were offered in Chapter 5, Numerical Calculation of Terrain Correction Within the Bouguer Anomaly Evaluation (i.e., the use of interpolated instead of measured heights for within the nearest zone), but this question remains open.
- Should we use the ellipsoidal approximation for outer and distant zones during the topographical effect evaluation? Or is it enough to

stay with the actual spherical approximation (with desired accuracy for different kinds of surveys)?

- Which of the distant effects (ignored in the standard Bouguer anomaly evaluation) play an important role in the interpretation of large scale (or global) geology? Distant relief effects (Mikuška et al., 2006) are relatively well known and also analyzed by different authors (e.g., also here in Chapter 2, The Physical Meaning of Bouguer Anomalies—General Aspects Revisited), but there still remain other distant effects, which should be analyzed in more detail in the future (e.g., Szwillus et al., 2016).
- Do we fully and properly understand the role of the Bouguer reduction (correction) density in Bouguer anomaly evaluation? Is the commonly used value of 2670 kg/m^3 really some average density of upper crust? Hinze (2003) gives us some thoughts about the historical reasons for this choice, but the topic is deeper and much more interesting....

As we have mentioned in the Introduction, the compilation of this small monograph was strongly motivated by the meeting of several experts at the workshop: "Bouguer anomaly—what kind of puzzle it is?" held in Bratislava in 2014. Perhaps, the time has come for a follow-up meeting, where these unanswered topics and new problems could be discussed and pursued. It may be useful for this next meeting to be covered by one of the professional associations such as European Association of Geoscientists and Engineers or Society of Exploration Geophysicists.

The editors of this monograph would like to thank once more all individuals and institutions who have contributed to its publication.

REFERENCES

Cella, F., 2015. GTeC – a versatile MATLAB tool for a detailed computation of the terrain correction and Bouguer gravity anomalies. Comput. Geosci. 84, 72–85.

Fairhead, J.D., 2016. Advances in Gravity and Magnetic Processing and Interpretation. EAGE Publications bv, DB Houten, e-book, 352 p.

Helmert, F.R., 1884. Die mathematischen und physikalischen Theorien der Höheren Geodäsie, Teil II. Teubner, Leipzig, (in German), 610 p.

Hinze, W.J., 2003. Short note: Bouguer reduction density, why 2.67? Geophysics 68, 1559–1560.

Hinze, W.J., Aiken, C., Brozena, J., Coakley, B., Dater, D., Flanagan, G., et al., 2005. New standards for reducing gravity data: the North American gravity database. Geophysics 70, J25–J32.

Hinze, W.J., von Frese, R.R.B., Saad, A.H., 2013. Gravity and Magnetic Exploration. Cambridge University Press, New York, 512 p.

LaFehr, T., Nabighian, M.N., 2012. Fundamentals of Gravity Exploration. SEG Tulsa, 218 p.

Li, Y., Krahenbuhl, R., 2015. Gravity and Magnetic Methods in Mineral and Oil&Gas Exploration and Production. EAGE Publications bv, DB Houten, 155 p.

Mikuška, J., Pašteka, R., Marušiak, I., 2006. Estimation of distant relief effect in gravimetry. Geophysics 71 (6), 59–69.

Szwillus, W., Ebbing, J., Holzrichter, N., 2016. Importance of far-field Topographic and Isostatic corrections for regional density modelling. Geoph. J. Int 207 (1), 274–287. Available from: http://dx.doi.org/10.1093/gji/ggw270.

Tsoulis, D., 2003. Terrain modeling in forward gravimetric problems: a case study on local terrain effects. J. Appl. Geophys. 54, 145–160.

Printed in the United States
By Bookmasters